The Electronic Bookstall

THE ELECTRONIC BOOKSTALL

Push-button publishing on videotex

Rex Winsbury

A critical study of Prestel in Britain:
the first two years of the first live videotex project

International Institute of Communications

First published 1979

International Institute of Communications
Tavistock House East, Tavistock Square, London WC1.
ISBN: 0 904776 10 7

Typeset by Red Lion Setters, 22 Brownlow Mews, London WC1
Printed by Penshurst Press, Buckingham House, Longfield Road
Tunbridge Wells, Kent
Design by Fiona MacGregor

Contents

REX WINSBURY is Editorial and Marketing Director of Fintel Ltd. Fintel is a joint subsidiary of the *Financial Times* newspaper of London and of Extel, which operates financial news, company information, printing, advertising and computer services in London. Fintel is the largest provider of business information to Prestel, the UK public videotex system.

Rex Winsbury was formerly Features Editor of the *Financial Times*, and is a graduate of Oxford University. In 1975-76 he was Thompson Fellow in Mass Media Studies at the University of Strathclyde. He then joined the management of the *Financial Times*, specialising in the application of advanced technology to newspapers.

He is author of two previous studies: *New Technology and the Press*, for the British Royal Commission on the Press (1975) and *New Technology and the Journalist* (1976), for the Thomson Foundation. He is also author of two non-fiction works for children, and is editor of a forthcoming McGraw-Hill book on videotex systems. He has lectured widely on Prestel.

Introduction

Videotex may be a revolutionary new medium of communications that will have significant social effects on the style and quality of life, as some protagonists claim. This study has a more modest scope. It is to analyse in some depth what has happened so far with Prestel, the British videotex service which is the most advanced of the many services that are now springing up round the world. (A note and glossary of the terminology is added to this Introduction). By "most advanced" I mean in an operating sense, as Prestel has about two years of practical experience to learn from, rather than in a technical sense, where other national systems may have advantages.

The objective of the study is to help the development of videotex internationally, by drawing attention to both the good and the bad points about the UK system, and to the arguments and debates that are taking place, in the belief that most of them will be relevant to other national systems, however diverse. Most importantly, it may help others to avoid the mistakes, and to build upon the strengths of the British approach. There is no sense in different countries continually "re-inventing the wheel". For the benefit of Prestel specialists, one might risk a metaphor and say that there is a danger of many countries "not seeing the wood for the tree structure", since the notion of Prestel as a "tree" of information is already a deeply embedded fallacy, one of several that this study tries to dispel.

Although the chapters range widely over the videotex territory in an effort to pick up at least an echo of all the main debates, there are two main centres of attention, two broad themes, around which the discussion revolves.

One is the practical business of creating information on videotex — how do you edit, what are the constraints, how do you take best advantage of the visual and technical characteristics of the system, what are the ethical and competitive rules, who do you employ? The other is the equally practical but more general questions of the relationship between videotex and traditional publishing media — if indeed the word "publish" is appropriate at all (I believe that it is, though not in a simple sense). How will the traditional media be affected, what are the differences of videotex, do traditional conceptions of publishing transfer easily to videotex, who will the customer be, what organisations, old or new, should "publish" through videotex?

Because I believe in videotex, and work in and for the medium, this study does not represent that school of thought (widespread if fragmented) that regards the whole development as a total waste of time. Anyone with his job at stake (as mine is) would be foolish to ignore the bad points. But only by working in the medium, and by being to that extent committed to it, can one get to know the arguments in the detail and subtlety that is, I hope, displayed in this study. It is fair to add that at this stage of the game the main problem is often to convince people that there *is* a problem about, say, editing for videotex, or about how you market it: convincing them of a particular solution is the second stage.

This study will therefore have failed if it does not do two things. One is to convince people that the creation of successful videotex systems is considerably more complicated than appears at first sight. The second is that the whole subject is deeply fascinating and, in the end, immensely worthwhile, both in the satisfaction gained from successful communication of any sort, and in the commercial profits and public service benefits which some videotex participants, but assuredly not all, will gain from it in the end.

The terminology

Unfortunately, there is no internationally accepted terminology to describe what is going on. The British systems and its participants have always used the terms "viewdata" and "teletext"; but more recently the International Telecommunication Union, which is the forum for the standardisation of these matters, has provisionally adopted the word videotex, and so this study uses the word videotex throughout. The terminology can be explained as follows:

Videotex is the generic name used for electronic systems that use a modified TV set to display computer-based information. Interactive systems using, typically, the TV set and telephone line, are called telephone-based or interactive videotex. Broadcast services are called broadcast videotex.

Viewdata is the UK term for interactive videotex.

Teletext is the UK generic name for broadcast videotex services.

Prestel is the proprietary "brand name" of the interactive telephone-based videotex (viewdata) service of the British Post Office.

Ceefax is the broadcast videotex (teletext) service run by the British Broadcasting Corporation.

Oracle is the broadcast videotex (teletext) service run by the independent television companies in the UK.

PTT means the national postal, telegraph and telecommunications authority.

Information Provider (IP for short) is the accepted term for a publisher on Prestel.

Videotex sits at the cross-over point between two rival jargons, two competing private languages. One, the computer language, speaks of a database maintained by manual or automatic update, with operators keying in, or bulk input by magnetic tape transfer or computer-to-computer handshake, with data reformatted to fit the Prestel display mode and generate the indexes.

The other, the publishing language, speaks of a magazine or information service published on videotex, with news, advice and useful figures prepared and kept up-to-date by editors and presented on Prestel pages with suitable lay-out and cross-references, using editing keyboards.

There are a host of associations and assumptions that go with each language. Neither set is wholly appropriate to videotex, but both are partially appropriate. That is one measure of videotex's novelty and challenge. Both languages are used here, inevitably if sometimes confusingly.

1 Prestel in principle — the common carrier philosophy

The most fundamental principle of the UK's Prestel Service is that the British Post Office runs it as a common carrier. It was an unusual decision. The Post Office has never previously operated as a common carrier — on the contrary, British law allows no such notion. And in the early years of Prestel, the Post Office wanted to be the editor or coordinator of the Prestel database. But after some arguments, in which several publishers said they would not cooperate with the Post Office on that basis, the Post Office decided, for the first time ever, to act as a common carrier. That decision is essential to an understanding of most of what follows.

The common carrier principle means that anyone can become an Information Provider, that is, buy space on the Prestel computer (or computers) and say whatever he or she wants (subject only to the law) to whoever he wants, at whatever price he wants. The Post Office exercises no control over content, quantity, quality, relevance, style, price (except within broad system limits) or the frequency with which information is changed; or indeed on how it is sold or marketed to the customers. The role of the Post Office is that of a neutral operator of the technology through which other parties communicate with each other. Hence the term "common carrier". Because that notion does not exist in British law (unlike, say, in the USA), various contortions have been necessary to arrive at a de facto common carrier position, including a clause in the contract between the Post Office and Information Providers by which the IP agrees to indemnify the Post Office if the Post Office is sued (as it can be) and loses the case. The matter has still not been entirely cleared up, but the present Post Office managers of Prestel are totally committed to it as a doctrine, and it is greatly to their credit that they showed flexibility in coming round to this doctrine during the early arguments.

The common carrier principle is at once the greatest strength of Prestel, its most exciting and innovative feature; and its gravest source of weakness. The weaknesses are mainly discussed in other chapters, notably those on the Code of Conduct (the problems of maintaining advertising standards and ethics) and on Marketing, where the problems of providing a satisfying information service and selling it to the user are discussed. Here the virtues are set out, and they are shining; but they must be evaluated against the disadvantages discussed in the other chapters.

The common carrier principle transforms Prestel from being just another computer system into being a genuine new publishing medium. By enabling (in theory if not yet in practice) any information supplier, big or small, new or established, to communicate information of whatever kind he chooses to any audience, big or small, that he chooses, it offers a range of choice, a freedom of expression, a flexibility, and a responsiveness between suppliers and users, that is unique in electronic media, and conceptually at least is a rival or complement to the massively variegated output of traditional printed publications. What is more, for the first time it creates a genuine market-place for information, quite unlike either broadcasting or existing computer systems. Many suppliers can offer their information wares to many users at a range of variable market prices, and either succeed commercially, or not.

It is here that the major innovation occurs; in the market-place not in the technology itself, which is by contemporary standards rather primitive. The innovation lies rather in this idea of a market-place for information, in which quite small lumps of information — the individual Prestel page of, at most, 130 words — can be individually priced, can have its price changed, can have its content changed, can be leased to a third party, can appeal to only a few people or to millions, and be supplied by anyone from government to Joe Citizen.

While some may jib at the idea of associating information with a market-place, since it smacks of huckstering and commercialism, others quite evidently do sense the intellectual excitement and the commercial opportunity of this "unbundling" (to use a computer industry term) of computer-based information. Computer systems have traditionally been high price, block subscription, narrow audience, specialist media. With Prestel, they enter the mass communications era.

As a publishing medium in this "open house" sense the costs to the publisher are in some circumstances attractively cheap. The storage charge for putting material on to the Prestel computer is £4 per page per annum. Even taking into account the £4,000 per annum "entry fee", it means that for £5,000 a

200 page magazine or information service could be maintained. These sums are cheap compared to trying to print, say, a 3,000 word weekly newsletter and send it to subscribers. Admittedly, the labour costs associated with collecting and collating information and creating its presentation are far higher than people not in the information business are apt to imagine. Any substantial Information Provider to Prestel will tell you that his main costs are labour costs, not system costs. Nevertheless, the fact remains that the basic system costs (the equivalent of the printing and distribution costs in traditional publishing) are low to the publisher. This is the other essential ingredient in making Prestel (or any other videotex system) a "free market" in information, since we are now dealing in mass-marketable commodities; that is, individual low-priced pages, which practically anyone and everyone can afford, if they choose to do so.

There are also clear political and social overtones to the common carrier conception. Any means whereby A can talk direct to B and C and D may become a means for A to influence B and C and D; and in a free, democratic society, why not? But the fact is that many mass communications systems, even in democratic societies, are not run on that principle. Certainly, if the pure common carrier idea is to survive, then it needs as much political as commercial support.

The most likely line of criticism is that, however beautiful in theory, the common carrier policy just creates a mess. It leads to a jumble sale of shoddy second-hand goods, rather than to an Eastern bazaar full of gold and jewels and rich tapestries. With no editor or coordinator, there is no guarantee that the information is complete, up-to-date, systematic, or even on the system at all. Nor is accuracy guaranteed. "Caveat emptor" must be the cry in this as in any market-place. The problem might disappear when there is a huge population of Prestel receiver sets and unlimited supply of space on the Prestel computers. Then, it is suggested, the normal market mechanism will drive away the suppliers of shoddy goods, as suppliers of better goods enter the market and attract the customers. But we are far from that stage at present, and there are grave short and medium term disadvantages, to which certain partial solutions are being proposed.

There is no necessary connection between videotex and the common carrier principle. Many systems (see list at end) are run on quite different lines. And in Britain the shortage of

Prestel pages may always limit the application of the principles. But let no one doubt its intrinsic excitement, its journalistic appeal, its social and political innovation, and its close alliance with the notions of a free press and free communications. For all the objections that this study voices (and they are weighty) the vision is a stirring one. It is visions that stir people into action, and people that stir technologies into action. Not the other way round.

2 Prestel in principle — whose code of conduct?

Newspapers and television have codes of conduct which lay down what can and cannot be said in the news columns, what can and cannot be shown on the screen and, in much greater detail, what can and cannot be said in the advertisements. Codes vary from country to country, and between newspapers and TV, as do the methods of enforcement, ranging all the way from voluntary to statutory, from rule books to case law, from pre-vetting to post-facto complaints tribunals. What code should Prestel have — if indeed it should have one at all? The Post Office has decided not to regulate Prestel's contents. Should a special body be set up? Or should the Prestel community be allowed to regulate itself?

There is a strong argument that a new medium should not be fettered with a code of conduct before it has even started, and before the real problems about taste, truthfulness, etc, have emerged; let alone before the exact nature of advertising on the medium has been determined. But there are even stronger arguments for the opposite views. In today's climate of consumer rights and consumer protection, it would be foolhardy to ignore the public's expectation that there will be a code of conduct to protect them. Even more important is the near-certainty that if the Prestel community does not create its own code of conduct, someone else will step in and do it for them. That "someone", however well-intentioned, may neither understand nor like the medium, and may fetter it far more than would a self-imposed code. Given the number of bodies already entrenched in the regulation business, from the Independent Broadcasting Authority through to the Advertising Standards Authority and the Press Council, there is no lack of candidates for the work of outside regulation, and no lack of voices demanding a code.

What the Prestel industry has decided to create is a voluntary code of conduct which all Information Providers will be asked to sign. It covers certain editorial questions, like not charging money for index pages, not misleading the customer as to what he will find on the information pages, and not misleading him as to the topicality of the material. But the bulk of it covers advertising questions. Many of the principles have been borrowed wholesale from the various codes that regulate British broadcasting and the press, as might be expected. But the unique nature of Prestel can create problems when the two models diverge, as for example they do on the question of what companies can say about their profit record. The IBA's advertising code debars reference to profits on the grounds that when quoted in a TV commercial they can have a highly emotive and misleading effect. In newspapers, however, as in company reports themselves, the figures for profits are a normal and indeed necessary part of certain corporate advertisements. Interestingly an early draft of the proposed Prestel code of conduct took the broadcast view of profits, until business IPs pointed out that references to profits were a normal part of their business information service.

Such matters may be only teething troubles at the drafting stage. But there are more fundamental questions. The most thorny is perhaps cigarette advertising; here again, the printed press is more lenient. Cigarettes may be advertised in newspapers and magazines (though with a printed government health warning) but not on television. Should it be allowed on Prestel? And what if the tobacco companies do not buy "advertising" space on Prestel, but buy their own space direct (a possibility discussed on pp 57-58) to promote the use of their products?

We have here a double problem. When the rules in other media conflict, which (if either) should prevail on Prestel? And what anyway counts as being an advertisement on Prestel, such that the rule, whatever it is, applies at all?

No doubt such questions, once recognised, will generate an answer sooner or later. But there is a further layer of questions relating to the enforcement of the rules that are laid down. There is a school of thought which says that the Post Office should enforce the rules, and simply throw any offender off the system. The Post Office undoubtedly has the technical capacity to behave in this manner. But it has adopted, and is sticking to, its policy of being a common carrier, and insists that this means taking a neutral stance (within the law) towards any material, editorial or advertising. Information Providers also would fight the idea of the Post Office being a judge of public taste. There is a further difficulty, in recording for subsequent judgement pages which may be on the system for a short time — hours or even minutes. You cannot run off an archival magnetic tape of the entire database every hour, just in case there is a single offensive page.

A voluntary code, such as that worked out by the Association of Viewdata Information Providers, the industry's trade association, could be made a condition of membership of the Association, and breach of it could be made grounds for expulsion from AIP. But here the problem (which may not be too great in practice) is that no Information Provider is compelled to join the Association, and membership of it is not a condition of being a provider to the system. There is an argument that a tribunal of AVIP members should recommend to the Post Office that an offender should be barred from or thrown off the system. But that notion offends other people as being reminiscent of kangaroo courts. But some method of adjudication, and some sanction, seem necessary.

Many points in a code of conduct, such as those covering advertising of specific goods (alcohol, drugs), the reduced-price sale of goods, appeals to children, etc, should probably not present too many difficulties. But the newness of Prestel means that not all the rules are self-evident, or self-evidently necessary. It may even be that the call for regulation in the name of consumer protection will one day become so strong as to be yet another fundamental challenge to the very notion of a common carrier service. Whether matters will ever get that far, only time will tell. In the meantime, the most important thing is that the industry should be seen to be taking sensible steps to keep its own house in order, in the light of contemporary standards of consumer protection.

3 Who's who in Prestel — united we stand?

For its successful operation, Prestel depends on the coordinated activities of four separate groups. They are (1) The Information Providers who actually have the information; (2) the Post Office, which has the computer which stores the information, and has the telephone lines that transmit the information to the user; (3) the manufacturers who make the terminals by means of which the user gets at the information; (4) the television rental companies who, by and large, are the main agencies by which the sets are actually placed in the home or in the office. Through it all, it is the Information Providers who are trying to sell their products, information, to the user.

This circular flow is very like to the flow chart of commercial TV in the UK, with its pattern of programme production companies → the Independent Broadcasting Authority which owns the transmitters → TV set makers → TV set renters → users. But there are significant differences. There are over 100 Information Providers signed up at the moment, with no upper limit, as against 15 ITV programme contractors; and the Post Office is seeking to act purely as a "common carrier" rather than as a regulatory-cum-distribution authority like the IBA. The content and layout of Prestel information depends upon the IPs. Because there is no central editor or coordinator, the information on Prestel is basically chosen and categorised by IP, and not by subject, although there is a subject index superimposed upon the IP indexes.

The IPs by and large have developed a close relationship with the Post Office through a series of committees that have considered pricing, technical improvements, marketing and other matters, and there is an Association of Viewdata Information Providers which tries to coordinate IP policies. In terms of size, IPs range from those with 100 pages or less to those with over 5,000 pages, and inevitably it is these bigger ones who carry on most of the direct dialogue with the Post Office.

Given the "open house" policy of operating Prestel, the variety of organisations which have signed up as Information Providers is extremely wide. Only some of them are traditional publishers of newspapers, magazines, books. Many are organisations who sense an opportunity to provide their information direct to the public for the first time. Whether Prestel is a publishing medium for hard data or soft data or both (that is, for facts and figures, or for written text and literature, or for both) it throws the field open for all comers who have the necessary cash.

Thus, for train times we find on Prestel *both* ABC rail guides, a private publisher, *and* British Rail, the train operator. For motor car purchasing advice, we find *both* commercial magazine publishers that publish car magazines, *and* the Consumers Association. For official statistics on, say, the retail price index, we find *both* Fintel, a financial publisher, *and* the Central Office of Information on behalf of the government itself. For theatre reviews we have *both* a magazine group providing reviews culled from its normal publications *and* an independent journalist (or rather, him and his wife) writing direct for Prestel. Then there are banks, universities, retail groups, computer bureaux, tour and travel operators, airlines, research bureaux, libraries, and so on. The gamut of information is similarly large — jokes and quizzes, classified advertisements, economic data, company information, timetables, price and shopping guides, share prices, credit card information, horoscopes, games, weather forecasts, and so on.

It is an open question whether Prestel is suitable for all or only some of these types of information. Will people read Prestel as they read a newspaper or pamphlet? Or is the glare of the TV screen such that they get dazzled quickly and will only pick out a figure, address, or time that they want? Will people browse through Prestel, or parts of it, adding up their own personal "information package" on, say, buying a new estate car; or will it be a quick in-and-out to find the availability of PanAm standby seats to New York tomorrow? Such are the questions that only experience of the market will answer. But those answers will affect who stays on Prestel as an IP, who joins it in future, and what information they decide to put up (since editing material on or off the system is relatively easy and instantaneous).

Two groups of IPs need special mention. Some IPs restrict access to their information to a known group of people or organisations (called a "closed user group"). Thus the information can be of a more sensitive, confidential, or expensive type, and the IP can if he wishes charge a simple annual subscription for membership of the CUG (or some other method of payment) rather than the page-by-page tariff rates of the public service. There is a view that videotex systems will find their widest application in closed user groups, either run off the same central computer or computers or off their own in-house machine.

The other group of IPs are those who either cannot afford the basic Post Office fees for being on the system, or who do not have sufficient material to warrant paying them. These IPs go to a larger established IP (known as an "umbrella" IP) and buy from them both computer space and design and keying services. Thus the system has room for both the very small IPs with a few tens of pages, and very large ones with 5,000 to 10,000 pages (or more as the system develops).

If the IPs are a polyglot community, so also is the Prestel community at large. Many of its members have never before had dealings with each other. The newspaper IPs are strangers to the TV set manufacturers, as are the TV set manufacturers to the computer software houses. Not surprisingly, there has been a great deal for all sides to learn about the others, and a lot of room for misunderstanding. Some of the most important and problematic consequences of having such a polyglot community are discussed in the chapters on the common carrier policy, and on marketing. Here we will discuss two other problems that have arisen in the "gutter" or boundary area between the various major interest groups involved in Prestel. One is the gap between promise and performance on the part of the Post Office in terms of the speed with which Prestel has moved towards a full public service; and the other is the gap between promise and performance in terms of the supply of Prestel receivers on to the market.

The origins of Prestel go back to 1970/71. Many people were quickly attracted to the notion of a massive, if not quite universal, database of information available to every citizen, and many organisations declared an early interest in participating in it as Information Providers. In early 1978, many companies (my own included) took the formal step of setting up special Prestel departments, even indeed separate Prestel subsidiary companies, to embark on the job of creating material on Prestel in readiness for the market trial of the system and the subsequent public service. In January of 1978, the timetable laid down by the Post Office read: June to December 1978, market trial with

1,000 test users; January 1979, opening of the public service.

But June 1978 came and went, and it was not until October that the first test user sets began to be hooked on to the GEC 4080 computer which the Post Office is using. January 1979 also came and went, and the total of 1,000 was nowhere near. A modified date of March 1979 was given for the start of the public service — and, indeed, in March a public service of sorts did start, for domestic users only, in London only, with only low level advertising and promotion. The total of sets of all kinds only crept past 1,000 in the summer of 1979, and a third of those were in IP offices, showrooms etc. The very notion of a test period, with results that could be evaluated in time for a public service, was virtually abandoned. The dates for a full public service in London, and for the opening of Prestel computers in other cities like Birmingham, Manchester, Edinburgh, slipped back to autumn 1979 or early 1980. In short, the job of setting up a successful nation-wide computer-based information retrieval system, on the massive and unprecedented scale envisaged, simply took far longer than anticipated.

This may not be surprising in hindsight; nor is there any doubt that the Post Office will get it technically right, sooner or later. But the delays have had two unfortunate consequences. One is that some IPs have already spent a lot of money keeping special units in being against the day when Prestel would really go "live". That money may never be recovered, and the end of the drain is not yet, at the time of writing, in sight. Other IPs have simply kept a low profile, watching and waiting. All IPs are now understandably a little cautious about future prospects.

The other consequence has been that it has further exacerbated the suspicions of the TV set manufacturers that Prestel was going to be a failure. The manufacturers had already been disillusioned by the low market penetration of teletext sets, after all the claims that teletext itself would be a revolution in TV use; and they were determined not to get their fingers burnt twice. What has led to great complaints from Post Office and IPs alike is that the manufacturers did not, as far as can be ascertained, admit to this pessimism. Instead, they promised to manufacture the 1,000 sets in 1978 for the market trial, and forecast another 50,000 sets to be made during 1979 (at least, their trade association did). In the event, the 1,000 sets were not all delivered until mid-1979, if then, and at the time of writing there seems little likelihood of there being more than 5,000 sets on the market by the end of 1979; perhaps less than that. What is more, no one outside the manufacturing companies, and perhaps few inside, know what manufacturing plans they have for 1980 and beyond.

The manufacturers point to technical troubles in getting the sets right and reliable, and to recent closures of factories. They blame the chip suppliers for delays, and they accuse the Post Office of misleading everyone about the timetable. They are resentful about thinly veiled allegations that they have "let the side down". But the fact of the matter is that an IP cannot sell his information without being able to supply, or point to a supply of, a set. Similarly, the Post Office rationally decided that it could not heavily promote a service, even when its own part of it came right, unless and until the customer could be assured of having a set to receive it. Thus even allowing for the slippage in the Post Office timetable, but recognising the massive resources the PO is devoting to making sure that in the end Prestel does work, it is the supply of sets that is the most serious and most worrying factor.

For what it is worth, the author's opinion is that the British TV manufacturers have indeed "let the side down", that they have not only misled the other members of the Prestel family, despite being closely involved from the start, but have also seriously compromised an exciting British invention in which Britain had a lead of two or more years over other countries, and in which export and other trade opportunities were and are substantial.

4 Who's who in Prestel — the trade union dilemma

Trade union structure and attitudes differ widely from country to country, and union reactions to videotex developments may be similarly diverse. But in the British context interesting questions have arisen about the adaptation of old unions to new technological circumstances, and about the entry of new unions into new job areas. So far, the unions have been cautious but constructive. Whatever British unions may or may not have done elsewhere, only in isolated instances have they actually impeded the development of Prestel. There are those who fear that this flexibility may change, as the number of jobs grows, but in the meantime the main concern of the unions has been to establish their interest and involvement in this new field.

Basically, the unions face three questions. For the established printing and publishing industry unions, the problem is whether work at videotex terminals is to be allowed to cut across the traditional lines of demarcation. For the telecommunications unions, the problem is how far their role in installing and maintaining Post Office equipment is to be taken over by private, commercial engineers and maintenance men. For all of them, but especially the printing unions, there is the problem that material for Prestel is prepared in many different places besides the established publishing houses. The source may either have no union at all, or an established union of a quite different industry (for example, a union of the computer industry, or, as in one case, the shopworkers' union). The arguments of demarcation usually centre on the process of putting (keyboarding) text and graphics into the system. Keyboarding can encompass writing, editing and formatting. It is a crucial job.

In the UK, the National Graphical Association, the compositors' union, is seeking to retain a monopoly of keyboard work, as evident in the 1978/9 dispute at the London *Times* and several less famous occasions. There is an obvious temptation to try and extend that monopoly, if only for defensive reasons, to Prestel work. Indeed, the NGA has formally adopted such a policy, and has enforced it in certain places, notably certain newspaper publishers where the union is already strongly organised. But other unions have equally strong reasons for having access to the keyboards (notably, the NUJ and some NATSOPA sections). If the NGA or any other union takes too hard a line, and tries to take on too much, and impose conditions that are too strict, it may drive Prestel operations into other sectors where other unions hold sway.

The nub of the matter is that the distinction between the writing (and graphics) of text and the setting of it in type, which is still deeply embedded in the European, though not the US, publishing industry, simply does not apply to videotex. There are some instances where the distinction has been imported from traditional publishing and forced upon the new medium, in yet another instance of trying to put new wine into old bottles. But such lazy thinking should be resisted. Intrinsically, the videotex writer will use the keyboard to create his material, if and when he wants to, given the severe constraints on his mode of expression of the small screen format: the librarian will use the keyboard to create his reference files; the advertising seller to create his advertisement; and so on.

Of course, it is *possible* to do all this work on paper and then give it to some other person who will use a keyboard to input to the computer. Indeed, some people for some purposes may find it easier and cheaper to do so; while others have already been forced to do so by union constraints. But it is not natural to do so, as a universal rule; it would certainly make the operation much more expensive if it were a universal rule, and the final result would certainly look worse, because there is no way in which the keyboard operator can be freed of the necessity of making choices and decisions as he or she actually enters material: the format is too constricted and demanding. In other words there is no exact analogy with pure typesetting. Whoever sits at the keyboard is bound to make editing decisions in many or most cases, quite apart from the all-important routing and indexing which is also a fundamental editing task.

Recognising this, as well as the inherent desirability of a writer being able to write direct on to his medium (as on to paper, via a typewriter) the main journalists' union in the UK, the National Union of Journalists, has adopted a policy which declares the right of its members to use the keyboards for editing as a basic right. The clerical union, NATSOPA, which organises advertising and library staff amongst others, has also laid its claim to access to the keyboards for its purposes.

Clearly there are the seeds here of some nasty inter-union

wrangles. So far, none has occurred, and one must hope that if they do threaten then solution will be found by inter-union discussion. The press is familiar with such arrangements (from the *Times* and other disputes). But videotex is not the press. Newspaper publishing has accumulated many conventions and routines and, in the case of the national press and most magazines, virtually has to be carried on in London. Neither of these constraints apply to Prestel. As a rough estimate, three-quarters of the material on Prestel comes from outside the traditional publishing industry. And editing and input for Prestel can be done anywhere, and by anyone with a suitable terminal and an accredited editing passnumber. Whether Prestel in the UK will ever form a coherent single industry may be doubted. There are already several differences between the Prestel service and its cousins Ceefax and Oracle, which have yet another set of system operators and unions. And even they are not uniform. At the BBC Ceefax service the editors work directly on to the keyboards, but at the ITV Oracle service they do not. There is a separate TV union that handles keyboard input. Thus does the world repeat itself.

On the engineering side, matters are no clearer, even in outline. The Post Office Engineering Union wants to retain as much work for its members as it can, including the installation of Prestel sets and their associated sockets. Whether it can maintain this position when Prestel sets are selling by the million, must be doubted. Like the NGA, the POEU can make life difficult if it wishes, without being master of its own fate.

In essence, we have here the familiar symptoms of existing institutions, the trade unions, when under stress from fast changing technology. The challenge is for them to react more constructively in an area of completely new activity like videotex than they have, in the UK at least, when the technology is applied to old areas of activity. So far, the omens, though mixed, are not on balance unfavourable.

5 Prestel in practice — editing the database

The most distinctive and most unfamiliar characteristic of Prestel and other interactive videotex systems is the structure of the database. Many people, especially those reared in the world of newspapers and magazines, see videotex as just a sequence of pages. Even those familiar with broadcast teletext tend to see interactive videotex as just a different way of displaying a larger number of screen pages. Yet the design of the database structure is the most fundamental activity involved in setting up a videotex information service. Without a good database design, the information service, whatever its information content, is dull, lifeless, hard if not impossible to use and, perhaps most importantly, fails to exploit the distinctive features of interactive videotex to create a distinctive product.

Almost as bad, there is a half-way house of computer and other experts who understand that these videotex services have a structure, but have interpreted that structure, on the analogy of other structures of computerised information systems, as a "tree structure". Articles and lectures abound in references to the tree structure. There is no tree structure: it does not exist, except here and there by choice. It is a pernicious concept that is almost as bad as having no conception at all — maybe even worse, since with no preconception the mind is at least free to grasp the reality.

The reality, and indeed the basic peculiarity of all interactive videotex systems, is that any page (a page being a single screenful) can have attached to it up to 10 automatic links or routings to any other page (as with most standard features, there are exceptions even to this generalisation in, for example, the Philips private videotex system). Thus the user can be told, when looking at page 10, that by pressing 1 on his keypad he will be taken to page 531078542 (or rather, to the information on page 531078542, since the user is not interested in the page number as such), and that by pressing 2 he will be taken to page 24807 (or the information on it), and so on. Only slightly less important as a peculiarity is that the user can get to any page at any time by simply keying the page number. Thus if he knows that the information he wants is on page 24807, he keys 24807, plus two control keys, and there he is.

It is up to you, as the editor or manager of the information service you are offering to the user, to make what you will of these two characteristics. Whenever a page of information is first created on Prestel, as on videotex systems generally, the editor working on it, or operator, call him or her what you will, first has to fill in a "shadow page" of instructions that the computer attaches to the page of information. This shadow page is the permanent companion to the information page, although invisible to the user. It is, as it were, the dark side of the moon. On this shadow page are entered the page number, page price, whether it is for general public access or restricted access only, and finally the routings or links for the information page, from 0 to 9. The editor may fill in all, some, or none of these, at his choice. But his choice it is, and the effect of his repeated choices, page by page, is to create a structure to the information, that is external to the information itself, but whose objective is (or should be) to aid the user to find what he wants.

Each page on Prestel has a unique page number, from 0 to 999,999,999, and in theory each of these has sub-pages, or frames, denoted by the letters of the alphabet "a" through to "z". Hence the theoretical capacity is 26 times 999,999,999. In practice, the first Prestel computers have a capacity of 250,000 pages, of which about 200,000 are available to Information Providers, and about 150,000 of which were in use by summer 1979. The way that Prestel is organised in practice is that the Post Office itself (ie, the system operator) creates a series of master indexes to the total database, beginning on page 0: and the individual IPs, of which there are over 100, are given (or rather leased for money) three digit numbers as their individual master numbers. Hence Fintel starts on page 248, the Birmingham *Post and Mail* on 202, American Express on 269, and so on. Three digit numbers are Prestel's status symbols: without one, you are known, a trifle derogatorily, as a sub-IP, since you have to use pages provided by someone else who has got a three digit entry point. Given the present shortage of pages, there has in fact developed a sort of black or secondary market in spare pages.

All the information provided by a particular IP sits "underneath" his three digit number, and in practice most IPs have tended to divide pages between index pages, which tell the user what he will find and the numbers to press to get it, and information pages, which are the user's final goal. There is no technical reason why this distinction has to be made, since, as we have seen, every page can have both onward routings and information. But it is a convenient distinction. One problem however that the creators of large databases have quickly found (those, say, of 5,000 pages or more) is that the larger the database, the greater the number of index pages as a proportion of the total, since the user may begin his inquiry in a number of different ways, and may wish to pursue it in a number of different ways. For example, cinemas or movie houses may have to be indexed alphabetically; by location; by price range; and by hours of showing; and the films or movies showing may have to be index alphabetically, by type (horror, cowboy, art etc), and by date of release (new or old) or even by country of origin.

Unless you cater for the many possible motivations of the inquirer, you will quickly lose him, or not get him in the first place, since he cannot easily find what he wants. It is not uncommon to end up with as many index pages as information pages; or indeed more. Now it is becoming increasingly rare to charge for index pages — how can you charge the user for discovering that what he wants is not there? As a result, half your pages may be carrying all your costs as an IP (a point taken up more fully in the chapter on costs and revenues).

As one answer to this problem, some IPs have consciously tried to abolish the distinction between index pages and information pages, as they are technically free to do, and to give information on all or most pages, simply varying the amount according to the number of routings that have to be given as well. That way, all or most pages can be charged for, and the user is spared the often tedious business of wading through large numbers of index pages before he gets to something interesting.

The disadvantage is that the Prestel page is so small. It has only 960 characters, of which only 858 are in practice usable, given that the top and bottom lines are reserved for system use and that the colour coding at the beginning of each line, except for white, takes up a character space. So even with only 1 or 2 onward routes displayed and described (eg, Press 1 for horror movies) let alone 4 or 5, the amount of space left for actual information gets increasingly constricted. If you are trying to give value for money, especially on high-priced pages aimed at the business market, you cannot give over half the page to non-information content. Each IP must find his own balance between these factors — as long as it is realised that there is a choice, or range of choices.

At this point, it is necessary to step back and ask, what is the overall structure of all these index pages, not to mention information pages, that you as editor or manager of a substantial database are creating? After all, neither the user nor indeed the editor ever sees all the pages at once, or even at one session. It is not like flicking through the pages of a 40 page or even 100 page newspaper, where you can see the overall structure of the product almost at a glance. Unless there is a consciously designed architecture, preferably expressed on paper, pages get lost, the editor does not know where to put anything, let alone where to direct the user, and the user does not know where to find anything. This is why, incidentally, it is not usually possible to generate either random or even sequential page numbers on a computer-to-computer transfer of data to Prestel. The need for an explicit structure makes it important that many sub-numbers remain the same (eg, the page number of the main local cinema, which customers get used to) even though the quantity of information may vary (e.g. how many movies, of what types, are showing there this month).

As we have said, computer people, familiar with the notion of a tree structure, have also seen videotex as a tree structure, or rather an inverted tree structure. They see it, in its crudest form, as a system in which the user is first offered an index with 10 general choices and, having made his first decision (eg, business information), is then offered a second index, a little less general, and makes a second choice (eg, market prices). He is then offered a third index, and makes a third choice (eg, share prices). A consequence of the way that Prestel is organised, and therefore a peculiarity of it, is that the user is then offered yet another index, showing the various organisations providing share price information, and he makes a choice (eg, the Stock Exchange). He is then confronted with the indexes of the Stock Exchange material itself, created by the Stock Exchange: these could be, first, a list of industry sectors, and second, an alphabetical list of company names, and only after making these further choices does the user eventually get his share price.

This quite realistic example (not exactly how the Stock Exchange in fact does it) shows that the user has to go through six indexes to get to the end information, and illustrates the "index page overhead". It is also the reason why there is a strong move to give the user the exact page number of the information itself, either by advertising or in the form of directories, rather like telephone directories, that he can look up, so that he can key the page number direct. It also raises the interesting question of whether the user will really be conscious of the transition from Post Office indices to IP indices, and whether he will really perceive the system as a single, Post Office information system, whatever the formal arrangements. This point is developed elsewhere in this study. But once a user has penetrated into a particular IP's database, one object of that IP's indexes and routing will be to keep that user inside his database, rather than have him wander off elsewhere.

But the question remains, is this picture of an inverted tree structure a valid one? Is it sufficiently valid to be made, as it is so often, the fundamental information structure of Prestel and of other interactive videotex systems? The supreme disadvantage of the tree concept in its pure form is that once the user has got to his destination page, what can he then do? Presumably, either ring off or start the whole laborious process all over again. He may be reluctant to do either, and anyway it could be disastrous to the economics of the system if the user only looked at one page each time he dialled in. This is known in the trade as the mineshaft syndrome.

So what do you do? What you do is to put on each and every information page details of further related choices that the user can make, and enable him to make them simply by pressing one key on his keypad. Another movie house? Key 1. Another type of movie? Key 2. Another area? Key 3.

Once you start on this line of development, you are in fact allowing the information itself to dictate, within broad limits, the external structure within which it is contained; and that is as it should be. But it also means that different types of information with different users may have different structures. At Fintel, for business information, we have identified various different structures, variously labelled cartwheel, ring of rings, lobster pot, chinese lantern, and other multi-dimensional models, as well as the simple tree. But the essential point is that they were created by editors working out for themselves how to present their information systematically, interestingly and dynamically on Prestel. Hence the inclusion of database design under the heading of editing; it is not, as so often presented, the computer manager's province, any more than the tree is anything more than one of a host of possible models or "architectures".

Prestel was conceived as a system simple to use, for the

ordinary customer. Talk of complicated data structures, as above, often draws the accusation that one has destroyed the basic simplicity of the system. But the truth rather is that to attain simplicity for the customer, a truly desirable aim, the editor has to evolve complex architectures of information whose complexity is virtually invisible to the customer but whose objective is to cater for the variety and unpredictability of the information needs of even the simplest human being. As with a physical building, the result of complex architectural design is usually a beautiful simplicity of appearance, and therein, we are all agreed, lies the supreme beauty of Prestel as well.

6 Prestel in practice — writing on the screen

If you turn the pages of Prestel, what do you see? Every medium has its own style. Its style is part of the way in which its audience identifies with it. Prestel is rapidly developing its own style, or range of styles. There are five elements. Two are traditional in any form of publishing: the lay-out of the pages, and the editing of the text and figures on the screen. But three are specific to Prestel. These are the easy availability of colour; the indexes or routing structures; and the pricing of the information.

The lay-out of a Prestel page is governed by the format of 22 usable lines, each with 40 characters. It's a very small page, about half the capacity of a normal VDU screen; and it will hold a maximum of about 130 words. But it has to be visually pleasing, or at least not so hard on the eye that you cannot read it. Each page must be self-contained, look well-organised, and have an obvious visual continuity with other pages of the same or related content. Each Information Provider develops his own house-style to cope with his own particular information, and the user will quickly realise that (for example) the Fintel magazine (or database, call it what you will) on Prestel *looks* quite different from, say, the Consumers' Association magazine.

In the same way, the editing demands that something useful be said on each page without fuss or flannel in a very short space. Prestel does not like long sentences or qualifications. Nor must you cram too many words on to the screen; otherwise it looks a mess. Each page has to have a single message, and a clear one at that. Two short paragraphs to a page, broken by a blank line, or two columns of figures to a page, are the most you will find on Fintel. Others try to get in more: perhaps, at times, too much, although a lot depends on whether the page is meant for the general or the specialist user.

Nevertheless, it comes down to what a newspaperman calls "tight subbing". The use of colour is totally different. It is a problem intrinsic to videotex. Seven colours are immediately available at the press of a button. The question is, does colour help to convey information better than black and white? There is a theory, which I do not believe, that business users of Prestel will prefer black and white TV sets, and domestic users will prefer colour. Experience suggests that most businessmen also want colour, though preferably a smaller screen than the present 24 inch models — say, a 14 inch set.

So how do you choose your colours? Every Information Provider makes his own choice. He could, if he wished, have every word and every figure on every page in a different colour, within the seven available (white, red, light blue, dark blue, magenta, yellow, green). Clearly, that would be absurd; but there are some absurd pages to be seen.

The ability to display in seven colours has caused some Information Providers to change colour almost from paragraph to paragraph, for no apparent reason: and/or to choose jarring or almost illegible combinations of colour to carry their message. In practice, the range of colour available is deceptive, since the deep colours — dark blue, magenta, red — are hard to read, and should rarely be used, apart from graphic applications. White is also rather glaring to read in quantity. There remain light blue, green and yellow as colours that are easy to read and easy on the eye. It will not surprise the computer community to learn that the Fintel database has fixed upon green as its basic text colour, with light blue for "key words", and yellow for page headlines or crossheads; leaving white for certain routing instructions, dates, etc.

In short, colour on Prestel must be seen in low key — comparatively few graphics, and comparatively few, restful colours for the text. Nevertheless colour is valuable, if used with discretion. Individual pages can be made more attractive to look at; their information content can be made clearer; a consistent house style can be given to an individual database; the division between the information content of a page and other (eg, routing) material can be made obvious; and last but not least colour can be used to make up for some of the limitations of the size of the screen. The use of colour can to some extent take over the use of white space in the lay-out of the printed page. In other words, colour can be used to differentiate one column from another, one subject from another, one category of information on the screen from another, thus tightening and improving the presentation.

Much more complicated is the index, or routing structure. Unlike a book, where you write the text first, and then add an index at the end, almost as an afterthought, with Prestel the design of the indices comes first. The indices form, in effect, the skeleton of Prestel, the bone structure which is fleshed out by the information. Bad design of the indices by the editor can render the information useless, confusing, or merely inaccessible. That not only annoys the user; it also costs him money which he will begrudge spending again.

For price is the fifth peculiarity of Prestel. Each small page has its own fee for looking at it, in addition to the telephone and computer access charges. With a newspaper, you pay one single sum for the whole lot: the same with a book. The BBC has a single licence fee per annum. But with Prestel, each separate nugget or page can have a price fixed individually by the editor, ranging theoretically from 0 pence to 50 pence, in practice from 0.5p to 15p depending mainly whether the page is meant for domestic or business use. So the pages have to be worth it. Not only should the indices not mislead, misroute or mislay; but the pages of actual information should be up-to-date, to-the-point, value-for-money. Pricing becomes part of the editing process.

Turning the pages of Prestel, you will therefore find five different sorts of page (everything on Prestel seems to go in fives). You will find pages to read, that is, text; pages of statistics, like share prices from the Stock Exchange or racing results from Sportdata; graphics, that is, pretty designs to make the thing look more attractive; index pages, those with the numbers 0 to 9; and what I call directory pages, those with lists of addresses, names, holidays, hotels, spare parts, consumer durables, cars to buy, all derived from lists that already exist elsewhere.

The bulk of the pages now on Prestel are of this directory type. Some argue that the small page size means that this is inevitable, and that Prestel will be like a gigantic reference book for people looking for very specific bits of succint information across a huge range of topics. Others argue that unless Prestel can be developed as an original publishing medium for comment, debate and argument, on social, political, financial or leisure topics, then it will never make its mark.

Several IPs have made a deliberate attempt to explore the possibilities of authors writing original material for Prestel. The text has to be written, or at least edited, into the special and highly constricted format of Prestel. One cannot let text spill over from one page to the next in random fashion; there is no way that anyone is going to read *War and Peace* on Prestel. The pages are too small, and the screen has too much glare to read a lot of pages in a concentrated manner.

On the other hand, painting produced the miniature as well as the panoramic canvas; literature produced the eleven-syllable Japanese poem, the Haiku, as well as *War and Peace*. Not that there is any poetry yet on Prestel, though I see no reason why not. But there are points of view on almost anything that do not need a million pages to propagate.

As to graphics pages, the fact that Prestel offers colour graphics at all is of course itself a distinct feature of the system. But it must be recognised that Prestel graphics are, at the present stage of technical development, very crude in design terms, being made up of comparatively few building blocks of colour in relation to the total screen. Unless and until they are refined to the point of being able to show, say, accurate road routes for motorists, the graphics will be pure decoration — and people will not for long go on paying real money for pure decoration. Graphics will have a limited use for some years to come — limited perhaps to title pages of individual databases and to advertising applications.

One of the uncertainties of videotex is the extent to which the different national systems will be able to show graphics, and how much users will be prepared to pay for better graphics. Prestel is a relatively simple system and its graphics are crude. The Canadian Telidon system can provide curves and other devices; but it is more sophisticated and more expensive.

The sources from which Prestel takes its information, and the relationship between those sources and what appears on the Prestel screen, are similarly various. There is, as mentioned, original material, specially written. But there are four other types. There are articles that have already been published in some magazine, say, the *Investors Chronicle* or *Which?*, and are then adapted for display on Prestel. You will find these in the Fintel or in the Consumers' Association magazines on Prestel.

Then there are articles which are presented in very brief abstract form, as in the industry and company reference section of Fintel, where each entry is a two or three line summary of a much longer *Financial Times* story. The lists or directories that I have already referred to are usually versions of some existing paper publication (eg, *ABC Rail Guide*, or Thomas Cook's holiday brochure). Finally, there are the news services, not all yet fully running, from such IPs as Reuters, the Press Association,

Fintel, Viewtel (Birmingham Post and Mail). These use Prestel simply as an alternative means of getting the news to the public. One interesting question is whether people with full videotex receivers will tend to get their news off Ceefax and Oracle, where it is free, and only use Prestel for the more complex information which Prestel alone has the capacity (at least at present) to cope with.

With these different sources, different lay-outs, different colours, different purposes, it is not surprising that the Prestel pages speak to the reader in different ways. The directory pages, those with lists, have the simple objective of providing facts, dates, numbers, prices. Other pages talk to you in a very conversational way, almost as if having a chat with you: this style is rather like a page in a tabloid newspaper.

Thirdly, there are advertising pages, discussed more fully in Chapter 8. We still do not know how much advertising there will be on Prestel, and quite what it will look like. Fourthly, there are computers talking to computers. Some IPs have their information on computer anyway, and simply (except that it is not all that simply) do a periodic machine-to-machine transfer. Whether the resulting Prestel pages are good or bad is perhaps a matter of taste. Some of the worst organised material to have appeared on Prestel, as well as some of the best, has been fed in computer-to-computer. On the other hand, in the long run it may be the cheapest and most efficient way to supply Prestel.

For all these reasons, the relationship between the publisher on Prestel and the reader/user is a novel and delicate one. When you tune in to Prestel and start pressing the numbers on your keypad, the IP may be speaking to you in several different ways. He may be saying, in effect, "I offer you *either* A *or* B *or* C to look at". That is the simplest version of the choices offered by the indexes. Or he may be saying "I offer you A *and* B *and* C", where he is inviting you to build up a cumulative picture of, say, a holiday in America or a British company.

Or he may be saying "I suggest you try A, *then* go to B, *then* go to C", where the IP is deliberately taking the initiative and guiding you by the hand round his "information stall". But equally, he may offer you a random collection of items for you to make your own sense of, in the form "A+L+P+B+Z." Or he may have a subtle approach, saying "Read A: if you need more detail, read B: if you need even more detail read C", so that different degrees of interest are satisfied in different ways.

For all these reasons, editing for Prestel is not like newspaper editing, and not like computer programming, but a novel and challenging mixture of disciplines drawn from both.

7 Marketing Prestel — divided we fall?

Interactive videotex, as organised in the UK, presents some intriguing marketing problems to which, as yet, only tentative solutions have been found. The problems arise from two basic characteristics of the institutional structure chosen, inevitably or otherwise, to house the medium.

As a result of the common carrier policy adopted by the British Post Office, the database on the supply side can and does look fragmented. The information is provided by many different sources, not necessarily to the same standard, and not necessarily making a coherent whole. Yet the user, in his office or at home, may well see Prestel as a single unified information service, simply because it comes through one vehicle, the TV set. Therefore the user's perception of the medium, and the system operators' perception of the medium, may be at odds. The second problem is that there are many different parties with a commercial interest in the consumer having a Prestel receiver. Each may carry out its own marketing, distinct and even at variance with the other, with the result that different sellers may address the same consumer with different or overlapping sales messages or calls, to the consumer's irritation and confusion. The marketing effort may therefore be dissipated and at least partly wasted.

There are the TV set manufacturers, who want to sell their own sets either to the consumer, or at least (in the UK) to the large TV rental companies. There are the TV rental companies, wanting to rent sets to the public, but not necessarily the sets of any particular TV manufacturer. There is the Post Office itself, the central system operator, with massive budgets and commitments, anxious to promote use of the system by everyone or anyone. Last but not least, there are the Information Providers, wanting to sell their information through the system; and even they are not a single body, but a miscellaneous group of over 100 different organisations with no common denominator except the fact of being on Prestel.

Yet the TV set makers and renters cannot sell or rent their sets unless there is information on the system which the consumer wants. The consumer is not in the end buying a piece of hardware; he or she is buying a means to get information. Yet the Information Providers cannot persuade people to buy their information unless they can also guarantee supply of a set, or be certain that the consumer already has one. The Post Office of course needs both information on its system and sets to hang on the end of it. This matrix of interdependence between otherwise independent (and powerful) parties is complex enough. But the result is that the IPs are probably confronted with a situation in which the consumer does not very consciously distinguish between their various offerings on Prestel, but judges their efforts as a whole, collectively, and simply says "Yes, I want Prestel", or "No, I don't". Thus marketing and promotional work done by one IP on his own behalf may, except in highly specialist areas, have limited value. Conversely, what one IP may or may not do, has a direct and inescapable impact on the effectiveness of the marketing and consumer appeal of another IP.

It would be tempting to pursue this argument to its logical conclusion and say that there is an almost Hegelian contradiction in the way that Prestel is organised. This contradiction would be between (a) the open-house common carrier policy adopted by the Post Office at the input end against its first instincts, but under severe pressure from publishing houses and other groups in a position to make a great political fuss about any attempt by the Post Office to set up, in effect, a rival publishing house, and (b) at the output end, the perception by the customer of Prestel as a single information service supplied through his or her TV set, with the consequent need for a unified marketing strategy to convince him or her to use the service. One could go further, and say that if the whole matter had been approached from the marketing point of view in the first place (as classical management theory would no doubt demand), then the whole institutional structure adopted in the UK should and could have been different.

The trouble is that an analysis on these lines, relevant and instructive as it may be for videotex systems in other countries that have yet to fix on an institutional structure, is largely of academic interest in the UK. The UK now has the structure it has, for good reasons, and it is almost certainly too late to change it (although not impossible, if the first attempt is seen to end in disaster). The immediate problem, being grappled with by various people (including the author, who therefore has something less than a neutral view of the matter) is how to make the best of what should be only a problem of the period of transition. For after all, when that golden day comes when every TV set is by definition a videotex set (as well perhaps as a home computer terminal, video disc display, and whatever else) then the matter will come round full circle, and it really may be simply a question of Information Providers trying to persuade people to look at their individual wares. The challenge is how to get from here to there, in something less than a millenium.

What in fact is happening is that a series of informal and sometimes overlapping ad hoc groupings are being formed, emphatically not to fix prices or market shares (or otherwise break the law) but to make the database itself look more easily usable by the customer, and to bring some co-ordination into the marketing.

On the information side, Information Providers are arranging to cross-reference between themselves by use of the routing mechanism on the system. Someone looking up a company in the Fintel database (say, ICI) is automatically cross-referred to the ICI share price provided by the Stock Exchange, and will be cross-referred also to the detailed financial analysis of ICI provided on Prestel by dataStream, a specialist computer bureau (and vice versa in each case). Similarly, sports results from Sportsdata cross-refer to sports records put up by *Guinness Book of Records*. IPs providing information about other countries are expected to start cross-referencing between themselves whenever they cover the same country. Several IPs in the consumer advice and information field have publicly announced their intention to co-ordinate their offerings on Prestel, and filling in gaps after discussion, and cross-referring. Thus there is a concerted move (without in any way infringing the rights of any party to do or change what they want) to modify the fragmentary nature of database. Here it is also worth noting that as long as the database is constricted in size and fully booked up, as it is now, so that new IPs cannot easily join and fill the gaps, then this activity is surely inevitable. Whether space on the database will ever be so freely available that any conceivable gaps in the information will be filled by natural market forces bringing in new IPs, must be doubted. This is a basic, intrinsic difference to traditional publishing, where a new magazine (if not newspaper) can always be started where a market gap is perceived.

The marketing of Prestel can be coordinated at many levels.

An Information Provider can combine with a particular set supplier, so that a single act of selling can take place. The customer who has agreed to buy the information service (which is what he or she really wants, whether it be the database as a whole or a special part of it) is then guaranteed a receiver set, either as part of the same package deal, or at least at a known price and delivery date.

Similarly, it would be surprising if joint advertising, promotion and direct mail shots did not take place between IPs interested in reaching the same group of consumers; just as some companies are already distributing printed directories and guides to the total database to all customers. This collaboration may be particularly active in business areas, as the businessman is the likeliest of all to perceive Prestel as a single information service, to be bought, or not, as a whole.

Again, it would be strange and indeed unhelpful if the advertising done by the Post Office (at present the major element in marketing) were not tied closely to the sales ambitions of groups of IPs, and of set manufacturers. After all, the Post Office (and indeed TV set suppliers) advertising is bound to feature the names of the more prominent IPs, and they are almost certain to demand a say in how their names are used. This opportunity means that they should have some view, if not a coherent policy, about how to market their information, and to whom.

The question of *how* may come down to two decisions. The IP first has to choose whether to do any marketing at all. For many IPs, the very notion that they have to do some independent marketing of their own, as opposed to just putting their information wares on show on the electronic bookstall, may come as a surprise, and an unwelcome and expensive one at that. Some IPs may decide to let others carry the burden. The IPs that do promote may do it singly, or with other IPs or, in some cases, as part of some wider marketing campaign that takes in other products of the same company (eg, other publications or information services).

To whom? may not be a problem if the answer is, everyone. The Meteorological Office, if it markets at all, presumably markets to everyone, although even they may decide to run specialist weather reports that require a special promotion. For business IPs in particular, there is a choice to be made. Should they market to a particular industry? A particular level of customer? A particular size of company? No doubt prices of agricultural feedstuffs need only be advertised to farmers. But what about share prices, or company information in general, or economic statistics? Will the user be a librarian or an information specialist? Will he be a dealer, someone who needs quickly to decide whether to buy or sell something — say, on the stock market or commodity market? Will it be the chief executive and other senior executives of the company? Or all of these?

Again, will the companies buying Prestel be large companies with large information budgets who are accustomed to buying information as a resource? Or are they so stuffed full of information already that there is no room for a new source? In which case, the real market for general business information on Prestel may turn out to be the small and medium sized businesses, the partnerships and professional people who are not large enough to have information specialists, libraries, or sophisticated computer information retrieval systems. To them Prestel may be a useful and innovative business tool, cheap to install, paid for mainly on a usage basis, easy to operate, and pitched at the level that they need. If this view is right, then it could and should colour the whole marketing approach of Prestel organisations of every kind.

Similarly, the so-called residential or domestic market may be really a series of sub-markets that need to be identified and attacked. In the end, like any other product or product range, Prestel information services demand standard marketing analysis and decisions; except that in this case it is not quite clear who can, and who should, make them to the best effect. In a sense, that is the central question to which this study addresses itself, and the author, for one, regrets that the present arrangements do not allow — perhaps, cannot allow — the creation of what might be called The Prestel Marketing Company to market the system as a whole on behalf of all its participants.

8 Advertising on Prestel — defining the undefinable

To computer professionals brought up on computer-based systems of information retrieval, the idea of advertising on such a system is an odd and foreign one. To publishing managers, of course, it is second nature. Advertising is, in one form or another, the life blood of most paper-based publishing. But the likelihood is that both parties will be surprised.

Prestel *is* an advertising medium and in this respect, as in so many others, is an odd kind of information retrieval system. Equally, its advertising may take some novel form (or forms), quite unlike the "conventional" styles that some conventional publishers are trying to promote at the moment. Furthermore, there is a latent tension, both ethical and commercial, between those who would ideally like to see advertising money and advertising pages kept rigidly separate from the main information content of the database and those who argue that, as with conventional publishing, the price to the average consumer will be far too high for commercial success and penetration unless the information service is not merely subsidised by advertising, but perhaps even taken over by it. They argue that sectors of the database should be "sponsored" by large companies much in the same way as sport is sponsored or, in the USA, some public TV programmes are sponsored.

We have two questions. What is the *nature* of advertising on Prestel? What is the *commercial role* of advertising within the overall economics of the medium? The peculiarities of advertising on Prestel arise from two factors basic to its operation. One is that it is a choice medium, in the sense that the user has to choose to look at each page, whatever its content. Because of the severely restricted size of each page, there is no question of using the standard print technique of scattering, on a single page, the advertisements among the articles in such a way that the reader sees them in passing, whether he likes it or not. Unlike printed publications, there is precious little link between general "page traffic figures" on Prestel and specific readership of the advertisements. The advertisements have to be looked at in their own right. This constraint tends to mean that they must have a high information content that attracts the reader for some practical purpose, rather than being some slogan about "Dizzo washes whiter" or "Bloggs — the store you can trust". In other words, even conceptually, the advertising takes its place as part of the general information content of the database, often indistinguishable in style and purpose from the non-advertising matter. This characteristic in itself should make one cautious about applying the conventional labels of "editorial" and "advertising".

The distinction totally disappears when one adds the second factor. Prestel is being run on an "open house" principle whereby anyone, whether a conventional publisher or otherwise, can buy space on the system and use it to speak direct to the public. In other words, organisations which in conventional publishing are advertisers, can go direct into Prestel and become publishers. Thus we find that American Express, Access credit card, Barclaycard, Comet discount stores, various travel and tour operators, airlines, and others, which in print-based media are advertisers are, on Prestel, Information Providers in their own right. A washing powder company can set up its own Prestel "guide to efficient washing", and the word advertisement would never appear anywhere, although of course the source of the material would be displayed.

It would be an exaggeration to conclude that the very conception of advertising, born in the print-based press and taken up by broadcasting, is quite inappropriate for videotex. It is not inherent, but it could be established. It is a question of organisation and control. For once there is central control of the information, the idea of central editorial responsibility re-emerges, and with that the idea of allocating certain space to other organisations who are not under this same central control, but are labelled as having paid money to have their message put there, as advertisers. But the common carrier concept, as applied to Prestel, virtually abolishes the distinction between advertising and other material except as a matter of stylistic choice.

To some, this may be a shocking notion: advertising is a thing to be distrusted, monitored, controlled. To others, the idea of the uncontrolled distribution of information may be anathema anyway, whether it masquerade as advertising or not. To yet others, it may be an exciting liberation from the conceptual straightjacket of existing media. But however you view it, it raises in acute form the question of what code of conduct, voluntary or otherwise, should apply to videotex services, as regards either their general information content or their so-called advertising content. This matter is discussed in detail elsewhere. But that discussion must be read in the light of the undoubted fact that material on Prestel, whatever may be true of other videotex systems, slides easily from one category to the other, and almost any conventional code of conduct (as is applied to commercial television in the UK, for example) is practically unworkable.

That said, there are some pages of what are called advertisements to be seen on Prestel, though at this stage not many of them, since the limited size of the audience so far has inhibited its appeal to advertisers. Broadly, one can discern three types of advertising: classified advertising, corporate advertising, and the "ad flash" technique to draw the readers' attention to a subsequent single page display advertisement.

Classified advertising, whose economic implications for the press are discussed elsewhere, looks fairly conventional. Its main characteristic is that the reader is required to make choices about what he wants, thus presupposing that he has a clear idea of what he wants in the first place. In other words, the classified ads are more deeply indexed than in conventional newspaper pages. The service has the advantage of giving the reader more precisely what he wants. Also, it leaves the ad on view for as long or as short a time as the advertiser wants; that is, the life of the advertisement is no longer constrained by the publication cycle of the newspaper or magazine, but only by the inherent "shelf life" of the ad itself. The disadvantage is that the reader is shut off from ads which, had he seen them by chance when looking for something else, might have sparked an interest. Classified advertising on Prestel is thus very different from classified advertising in print. Actually, it is very close to being pure information. It is called "advertising" only because, in newspapers, it clearly falls into that one of the only two available categories. Basically, the classified ad is just a statement that a certain item is for sale, and in the videotex systems of tomorrow (or the next day) it is quite possible that it will be seen as information rather than advertising. Furthermore, Mr or Mrs Adverage will put up his own information about his second-hand car for sale, rather than use an intermediary.

Corporate advertising is being developed by several companies as a "package" of frames which is sold to the corporate advertiser as a space within which he can put whatever information he wants about his company (press releases, financial results, company structure, etc). A company may thus appear twice inside an IP's database — in its own ad and in the file that the host IP has itself created about that company, and for which it assumes sole editorial responsibility. On Fintel, the corporate advertisement is clearly labelled as an advertisement, to distinguish its origin from that of the Fintel company file. But the interesting point is that a company wanting much more than the 100 frames, especially if it is a company confident of its ability to attract readers in its own right, can then simply move away from Fintel and become an Information Provider in its own right — in which case virtually the identical type of material appears on Prestel without the label of advertisement. Fintel has a dramatic case in point. American Express first bought the 100 frame advertising "package" inside the Fintel database, and then decided to exploit the medium in greater depth and to move their material off to their own 3 digit number (that is, became an Information Provider in their own right, although Fintel retained the contract to do the design and input work). The point here is that what began as a publisher-advertiser relationship became overnight an Information Provider-sub-contractor relationship.

It is simply a matter of where the material will gain maximum attention. If the answer is, inside someone else's database, then it may have to be labelled advertisement to distinguish it. If it is self-sufficient, then it will not be labelled as an ad. This criterion is far different from conventional advertising.

The ad-flash technique puts a one-line "taster" at the bottom of an information page (say, a page of news) to trigger off a choice by the reader to look at a subsequent full page advertisement. These full pages are sold to advertisers at a certain rate per day, week, or month, with extra fees for special contents such as graphics. The technique is an attempt to make advertising on Prestel approximate as closely as possible to advertising in the press and magazines. It is fair enough as far as it goes. But in the long run it may satisfy only those clients who do not have sufficient material, or sufficient funds, to become a direct supplier of information to Prestel. The big local department store, supermarket or furniture store will surely, sooner or later, puts its own information magazine on to its local Prestel computer. So in broad terms this type of advertisement may prove to be a temporary, though useful, phenomenon.

We could define advertising, whatever its format or appearance, as being any page for which the customer pays little or nothing for the information, because its costs are met by some other means. Is this advertising a desirable and possible means of advance for Prestel, or not? Much government information on Prestel may be free, as a public service to the citizenry. Travel operators put up their timetables and tour schedules for free, to attract travellers. Product lists and prices, from stores, manufacturers and mail order houses, will probably be free to the user. But what about company information, or political information or indeed any such information from a commercial publisher that has to earn a living?

Should this information be independent but expensive? Or sponsored by the advertiser, and so be cheap or even free? Should Prestel be like the BBC or should it resemble US television? Given the inherent high costs of creating information (many people do not realise how high that cost is), then the problem is a very real one. It will apply pari passu to many categories of information.

At root, the question is this — do the economics of public videotex systems require massive exterior subsidy, or can they be sustained on the basis of customer revenue? And, if desired and needed, what is to be the nature of that exterior subsidy, its rules, influence, extent, and expression? That, rather than simple labelling and formatting, is the interesting question about "advertising" on videotex.

9 Costs and revenues — four economic models

Will it make money? That is what everyone wants to know. But it is still too early to say. Partly because the market trial of Prestel has not been as well executed as hoped, there is little reliable evidence yet on the all important patterns of usage. The nub of Prestel economics is that you have to incur high fixed costs simply to put an information service or database on the system. After that, your costs do not increase at all, or hardly at all, as the number of users goes up. In newspaper terminology this is known as "first copy costs" — that is, the costs of getting the first copy of that night's newspaper rolling off the presses. It includes all fixed costs of newsgathering, the staff, the premises, composing and printing equipment, and vans. Thereafter, the only extra costs incurred for every extra copy of the newspaper printed are the extra ink and newsprint, and the transport costs.

With Prestel, and perhaps with most videotex systems, the matter is even more extreme. There are, by definition, no extra ink or newsprint costs, or transport costs. Such variable costs as derive from the number of customers are carried by the Post Office itself (computers, communication network) and by the customer himself (his modified TV set, bought or rented). Therefore, the IP on Prestel simply incurs a one-off cost to create his pages, and may lose it all if no one looks at those pages, or may make a fortune if the number of users, or rather page accesses, exceeds a certain critical figure that covers his initial costs. For if that critical figure is X, then every page access over X is pure profit. A mouthwatering prospect. But what is X?

Nobody knows for sure, simply because no values can yet be attached to the average pattern of usage of the average business or domestic customer. Crude numbers of Prestel receivers in the marketplace tell you very little. They may be used frequently, infrequently or not at all. Nor do we yet know the price sensitivity of the information displayed, or what, if any, competitive constraints there will be on pricing, either from rival IPs on Prestel, or from rival media. For any given page, we do not yet know at what price level the customer will refuse to look at it, or at what level someone else will step in (computer capacity permitting) and undercut.

However, some at least of the basic building blocks of costs and revenues have been fixed, or can now be estimated. On the basis of these and of certain hypotheses about patterns of usage, possible models can be drawn up of an IP balance sheet of the future. From these, some conclusions can be drawn about the receiver set population that will be required to sustain a viable Prestel industry. It is possible that the economics look quite different from the Post Office point of view, and different again from the TV set manufacturers' point of view. If so, it is yet another example of the problems that exist at the boundary or "gutter" between the three interest groups involved in Prestel.

Let us first deal with the tangible costs to the customer. These costs are quite complicated to work out, because the user of Prestel pays both the TV manufacturer, the Post Office, and the Information Provider for use of the system. To the TV manufacturer (or rental company) he pays at the moment up to £1,000 purchase price, or £30 per month rental. These figures are very variable, since at present sets are often placed at subsidised rates; and real prices will fall when volume production increases. To the Post Office he pays, first, the local telephone call charge, which at present is 3 pence for 2 minutes at peak rates, and 3 pence for 12 minutes at off-peak times. This money goes to the Post Office telecommunications department. Secondly, he pays a Prestel usage charge, which has varied so far but is currently expected to be 3 pence a minute at peak times and 1 pence a minute at off-peak times, with a minimum of 3p. This money goes to the Post Office's Prestel department, a separate profit centre.

Then, the user has to pay for the information itself. The system at present allows the IP to charge between 0 pence and 50 pence per page. The pricing pattern emerging from IPs is for advertisement and routing pages to be priced at 0 pence; for residential and consumer information to be priced at either ½ pence or, at most, 1 or 2 pence per page; and for business information to be priced at 5, 10 or 15 pence per page.

From these figures, we can make our first rough calculations. Assuming that it takes 15 seconds for a Prestel page to be displayed and read and that a user will look at four routing pages before reaching an information page, the average cost to a user for each page of useful information in the residential database, at off-peak rates, will be about 3p; and in the business database, at peak rates, about 12p.

The Information Provider also incurs a variety of costs for the risk or privilege of contributing to Prestel. First, the Post Office charges an "entry fee" of £4,000, and £4 per page per annum for each Prestel page booked. This means that an IP with a significant database — at least 1,000 pages — will have to spend £8,000 to join the Prestel club.

On top of this basic contribution is the cost of running the IP's database. It includes the cost of terminals and the use of the telephone line (most updating and inputting is currently done on-line) which goes to the rental companies and the Post Office. It also includes normal administrative costs, especially labour costs, which vary widely depending upon the size of the database, the frequency of up-date and editorial standards. The consensus among large IPs is that the overall cost of inputting and maintaining a page on Prestel, including overheads but not the Post Office charges, is between £20 and £30 a year.

Many people outside the information industry regard this as a surprisingly high figure. But information is by its nature a highly labour-intensive business, and so far the indications are that Prestel will be no exception. As we shall see, with a large database, labour costs can dwarf any other costs: there seems to be a rough correlation among IPs of one employee per 1,000 pages. Computer experts with data that is already computerised will plan to eliminate labour costs by machine-to-machine transfer. But the peculiarities of the structure and display of videotex systems make it much more difficult that computer experts at first realise to transfer anything but the most regular and simple data (ie, price lists or spare parts lists) by bulk update or magnetic tape transfer. Perhaps these problems will be overcome. But in the meantime, labour costs are high, higher than newcomers expect, and too high for comfort. The IP is faced at this stage with easily quantifiable costs, and almost unquantifiable income.

In the light of these figures, let us examine how the economics of Prestel look to each of the parties involved — the Post Office, the set makers, the IPs, the customers — and how they diverge. Then we will set out four hypothetical cost-revenue models for an Information Provider, with their implications for desirable set populations.

For the Post Office, penetration of the mass residential market is regarded as essential if Prestel is to generate the revenue that will make it profitable. An additional appeal is that it gives an opportunity for a large, under-utilised asset — the residential telephone network — to be more fully exploited. It may also offset falling employment in other areas of Post Office activity. However, the Post Office is primarily interested in ensuring that everyone with a telephone and a television set becomes a Prestel user. It is not particularly bothered whether the user buys a new TV set for £1,000 or an adapter for his old set at £100. In fact, the Post Office is seriously considering the idea of entering the adapter market in a big way.

For the television set manufacturers, the mass residential market is the only one that makes economic sense, and then only if the majority of that market buys or rents a new Prestel-receiving set. The British TV manufacturers see Prestel as the possible answer to the flattening out of colour TV sales now that the colour market is nearly saturated. But the economics of mass production dictate that sets must be sold in hundreds of thousands before the cost of laying down a new production line (perhaps about £15 million) can be justified.

To get the price down to the level of normal-set-plus-15% regarded as the level acceptable to buyers will itself require a market of considerable size. It is argued by the Post Office and the Information Providers that in order to create a mass market, the price of sets must be low to start with — in other words, that the manufacturers must start selling their sets now at mass production prices. The manufacturers, understandably, want more evidence that the mass market exists before taking such a step.

For the Information Provider, the choice of what market to aim for is not so clear-cut. As indicated, the IP will need an income of between £25 and £35 per frame to make participation in Prestel worthwhile. This income can be generated by a lot of users looking at the IP's pages with a low price per page, or a small number of users paying a high price per page. It can also be generated by a lot of users looking at a few pages frequently or a lot of pages occasionally. Also, the IP is not particularly concerned whether the user has a new TV set, an adapter with an old TV, or even an adapted VDU, or whether the user calls Prestel on congested business or under-used residential telephone lines — just as long as he calls.

Here then are clear areas of divergence in the economic interests of the various parties involved. How serious this divergence is, and whether it is more powerful than the general interest in making the system a success, remains to be seen. There is no reason in principle why the divergencies cannot be reconciled. What is sure is that the significance of the divergencies were not initially recognised, which should serve as an object lesson for other countries.

What of the Prestel buyers, the users? For the residential user, the nature of the Prestel system — paying for information both page-by-page and for connect time — will be novel and not necessarily welcome. Consumers at present buy their books, newspapers, televisions, on a fixed, lump-sum basis and, once bought, they can be used and re-used without incurring much extra cost. The pay-as-you-use system may inhibit the residential user because of the psychological effect of having an open-ended, hard-to-control expense, and the user may be reluctant to browse through the database because he will be conscious of a "meter" ticking up the pence as he goes from page to page.

As mentioned, the total average cost to the residential user for each access to a page of information will be about 3 pence, and it seems reasonable to assume that an average Prestel session will involve looking at two or three pages. The residential user will therefore be paying about 10p a time, a price that may be high enough to prevent Prestel becoming a regularly used source of information in the home. After all, a local newspaper, which contains a lot more immediately useful information, also costs about 10p, and even at that level there is evidence of sales being sensitive to price rises. The fact is that the consumer is not used to the idea of paying directly, per item, for information. As they stand, Prestel prices are not going to encourage him to change his habits. What is more, of the 3 pence average cost, 2 pence goes to the Post Office. This split makes it very hard for the residential Information Provider to test alternative pricing strategies. Indeed, it almost means that he has no control over his own pricing at all.

The major stumbling block for Prestel in its attempt to break into the residential market is the price of a Prestel TV set. As things are at present, with no manufacturer yet willing to take the plunge into mass production, sets are going to be expensive for quite a long time. To persuade a potential user that Prestel — intrinsically quite expensive — is worth a 50-75%

premium on top of the normal new set price is going to require ingenious marketing.

In the business sector however the picture is much more encouraging. First, a Prestel set is quite cheap in terms of existing VDUs and in terms of a company's capital budget. Secondly, companies are quite used to paying for business information, whether it be for credit ratings, market research, or a subscription to a news service. The business information market is not particularly price sensitive as long as the information is of good quality and, in terms of prices for business information, Prestel is very competitive. The question rather is, what business information will "sell" on Prestel?

The contrast between the residential and the business market is therefore almost total. The residential market will probably be acutely price sensitive, but the information may be fairly predictable — weather, entertainment, sports results, games, train times, consumer advice. The business market will not, within limits, be price sensitive; but given the wealth of information sources enjoyed by modern business, or at least by the larger organisations, it is not yet clear what business information will have appeal on Prestel, and to whom.

With these thoughts in mind, let us now construct four models, two for a residential IP, two for a business IP, of some possible patterns of costs and revenues. They are, as already emphasised, highly speculative. But in the absence of anything better, they are quite instructive.

Model A

A residential IP with 1,000 pages. He charges one pence per page for his information pages, and his index pages are free. His basic annual costs will be:

PO subscription	£4,000
Page booking charge	£4,000
Labour costs of one person full time plus half a supervisor, plus overheads, say	£16,000
Total	£24,000

Given that a third (say, 300) of the pages will be free index and title pages, he has 700 information pages to generate a median revenue of about £34 per page. Therefore, each information page, at one pence each, has to be looked at on average 3,400 times; or, allowing for the Post Office 5 per cent factoring charge and 20 per cent on revenue as a profit target, 4,250 times. Put another way, the IP needs a grand total of 2,400,000 page accesses at one pence a time, to cover direct costs, and another 25 per cent to cover Post Office deductions and profit, making 3 million in all.

Let us now make an assumption about the customer. We assume that he or his family are relatively heavy users, prepared to spend about £40 a year on Prestel (perhaps the equivalent of one extra telephone bill on top of the four quarterly ones he already gets). At an average cost of 3 pence per time, he looks at 1,330 pages of information in the year, or about 25 a week. Let us now make an assumption about the IP; namely, that he is a popular IP, perhaps a local newspaper with all the local events and classified advertising. Say, one third of these 1,320 accesses (440) go to this IP's pages. At 440 accesses per customer, the IP needs only 6,800 Prestel sets in use to reach 3 million total and to have a viable market. That may seem a low figure. But the IP's information, based on the local newspaper, would attract only sets in the locality or region of that newspaper. To achieve 6,800 sets locally, the national set population would have to be many times higher. It might have to be 50 or 100 times higher — say, a national population of 350,000 to 700,000 sets.

Model B

Let us now take the same IP with the same number of pages and the same costs (£24,000 a year), but charging only ½ pence per information page, on the grounds that for the residential customer charges have to be as low as possible. His revenues per information page are halved, so his total number of accesses has to double to 6 million. Let us now make a different assumption about the average customer. We'll assume that he is a light user (you may think intuitively that the 25 pages looked at a week in Model A is unlikely) and looks at only 3 pages a week (say, to check a cinema or train time), or 150 a year. Suppose further that your information, as the IP, is not such that the average person needs it very often. So of those 150, the IP gets only one in 30, or 5 over the year. Then simple mathematics demands a set population of 1.2 million. If you assume that the average person needs, for example, an unfamiliar train time even less often, say twice a year, then an IP with rail timetables needs three million sets; and so on. Given the range of information on Prestel, and therefore the competition for whatever amount the average customer is willing to spend on it, an instinctive judgement is that a million sets is rock bottom, and three million sets only just comfortable, for a residential market.

Model C

Let us now take a business IP with a big database of, say, 10,000 pages. His costs (given the quality of people needed to handle business information, and the fact that some information may have to be bought in) might be:

PO subscription	£4,000
Page booking charge	£40,000
Labour etc costs	£115,000
Overheads, extras	£40,000
Total, say,	£200,000

With a database that size, it is quite likely that half the pages will be index pages (if these proportions are surprising, please consult the chapter on database construction). Therefore 5,000 pages of information have to generate £40 each in revenue a year: adding 25 per cent for profit etc, that makes £50. If the average price per page is 10 pence, then that means an average of 500 accesses a year per page, or 2.5 million accesses over the entire database. Let us now make an assumption about our business user. Suppose he or she looks at ten pages per session, 3 times a week, or about 1,500 a year. Suppose that you are a popular IP among business users, and that one-third of those pages are yours. Then the picture is of the average business user spending about £200 a year on Prestel (including PO charges), of which £50 comes to you as the IP. In this case, the minimum market is 5,000 users — a low figure.

Model D

Let us now take the same IP with the same database and same costs, and same charge per page. But suppose that the average business customer only looks at 5 pages once a week, or 250 a year (the £200 in Model C does not sound high, but the frequency of use may). Further, suppose that you are not such a popular business IP, and that only one page in five is yours; ie, 50 a year. Then the desired minimum set population becomes 50,000 — ten times that in Model C, but still far short of the numbers involved in the residential market models.

Lastly, a few observations on these models. There is, one, a clear and fundamental difference between the set population required by the business IPs and that required by the residential IPs.

Two, if the problems of marketing to the residential user are anything like those described above, then three million sets may be a formidable target; even though it was, until recently at least, the official Post Office forecast for 1983.

Three, the business targets are much less. But the number of sets that would comprise a viable market for business IPs, even at the upper figure, is not a number sufficient to justify a set manufacturer laying down a special production line for Prestel sets especially if production is split among several manufacturers.

Conversely, if the business IPs have to rely on the residential market to create the volumes needed to get sets manufactured at all, they may have to wait too long for comfort.

Perhaps adapters, or special designs, or a new entrant into manufacture, are the answer to this dilemma for the business IP. What is for sure is that business IPs are praying for a solution, with little control over its discovery.

0 Costs and revenues — hoisting the umbrella

Around the primary activity of supplying information through the UK Prestel system, there has arisen a large and diversified secondary industry of organisations offering consultancy, training, conferences, research, software and hardware. Software and hardware are outside the scope of this chapter. But it is remarkable just how much peripheral activity Prestel has generated before the information service itself has really got going, let alone produced anything more than peanuts in terms of revenue.

Indeed, this may be the point. Information Providers have been conscious of the drain on cash of the longer-than-expected wait before the Prestel public service comes into full operation, and have been seeking to develop other sources of revenue to tide them over. The larger of them have therefore sought to diversify out of, and at the same time capitalise on, their adaptation of their own material on Prestel, by making money from the skills and knowledge acquired in so doing.

These organisations have acquired the name of "umbrella organisations". Most if not all of them offer to sub-lease pages on Prestel to other companies, and do the design and input work for those companies' pages. But most go further than that. They offer all or most of a package of public conferences, private seminars, intensive training courses, database design (house style and page design), business planning, supply of ready-made Prestel frames of information (for example, to videotex systems outside the UK), consultancy work, market research, and liaison with the Post Office and TV set suppliers, not to mention the writing of pamphlets and articles.

In close competition come specialist consultancy organisations who do not "own" pages on Prestel but are usually staffed by people who have left either the Post Office or one or other of the major IPs, and who offer a range of pure consultancy services; specialist conference organisers who also do not "own" pages on Prestel but who sense a market for commercial conferences on Prestel developments; and the computer consultants who have naturally been attracted into this area. It makes for a highly diversified and competitive secondary services market.

Some organisations have already dipped into and dipped out of, this secondary market. In so far as it is skill and knowledge that is for sale, it may take a while to shake out the wheat from the chaff. But the idea is now established that the development of Prestel (and perhaps of private videotex applications in the UK) may revolve around a limited number of "hot shops" who are basically very good at videotex. These "hot shops" will run information services on Prestel, advise others what to do and how to do it (and do it for them, if required, for a fee), and lecture the world, for a fee, about what is going on. There may be room, in the end, for at most half a dozen of these "hot shops".

In most ways, it is a very legitimate activity. There is an enormous groundswell of interest in Prestel to be satisfied; and there are many potential small Information Providers who do not want to have to "re-invent the wheel" every time, and who will willingly buy experience and/or contract out implementation, knowing that this is the best way to get value for money.

Also, there is virtually a declared policy by the British Post Office to encourage smaller Information Providers to go in under the umbrella of larger IPs — hence the origin of the term "umbrella" organisations. The Post Office has favoured umbrella IPs for two reasons. The practical reason is that the desirable pages on the first Prestel computer, some 200,000 of them, were all booked up well in advance, so that during 1979, or even earlier, the only way that a new IP could get on to the system was by sub-leasing pages from someone who already had a large block booking, and a few pages to spare. Indeed, something of a "black market" has developed in Prestel pages because of their scarcity, and it may get worse before expansion of the system eases the shortage. What is not known, at the time of writing, is when the Post Office may double the size of the database by adding another disc, and so end the rationing period.

But there is a second reason for the Post Office encouraging "umbrellas". It wishes to ensure quality of control. There is no central editor or supervisor of the material on Prestel. On the contrary, the Post Office has adopted a deliberate policy of not interfering in the contents of the database for any reason whatever. The disadvantage of this otherwise laudable policy is that Prestel can (and does) contain rubbish as well as high-quality data. But at least the umbrella organisations will contribute their own skills and presumably high standards to material they create for others, if only to ensure that their own reputation does not suffer. Equally, they will tend, by the weight of their presence, to set standards for the medium as a whole. The hope is that the good will drive out the bad, and not vice versa.

There is perhaps a danger of imposing uniformity on the database; and of precluding original ideas. But when the shortage of pages eases, this danger should be more theoretical than real. At the moment, the danger is the reverse — there is too much incomplete, out of date, badly presented, over-priced, badly routed, badly indexed, boring, visually displeasing, wastefully laid out and generally awful stuff. For the user in these early days, a turn-on could be an immediate turn-off.

That "umbrella organisations" become centres of excellence can be no bad thing. Also, the fact that they are competitive between themselves must benefit the service as a whole. The prices they are charging seem to be settling down to some recognisable pattern. In the early days, some wildly low prices were being quoted, before people quite realised the full labour costs involved in creating and maintaining Prestel pages. There still are some ridiculous prices quoted, but not so often. Clearly, there are distinct differences in operating costs between an organisation operating with full time staff in the centre of the capital city of London, and another offering similar services from a remote (though not in telecommunications terms) provincial town using local and/or part-time staff. Such a difference is bound to be reflected in different charges. But at least a recognisable market place, with prices, has emerged for videotex consultancy and input services, which is more than can yet be said for videotex information services.

11 Videotex and newspapers — telepress?

What we can tell the press about Prestel? Is it a threat, or a promise? Is it to be ignored, deplored, explored? The publishers' reactions to these questions have already been highly emotional, ranging from "We have nothing to fear" to "It will bury us". The reactions from the videotex community on the contrary are anodyne. For reasons that are essentially diplomatic, the party line is that the two media are complementary, and that everything is for the best in the best of all possible worlds.

All this is rhetorical rubbish. While it is too early to be dogmatic, one can with reasonable evidence declare a more sensible order of things. One, it will take a long time for the effect for videotex systems to be felt on newspapers. When it comes, the effect will be patchy; that is, felt in some areas and respects, but not in others. But effects there will be, forceful and unambiguous, which will not spell the doom of the newspaper but add to the pressures for further modification of its role in the electronic era.

The speed, direction and force of the impact on the press of videotex systems will depend on a range of factors governing its acceptability to the general and business publics. Seven main factors can be seen.

Legibility. People are not used to reading the TV screen, or indeed any other screen. Habits can die and grow slowly. It may take a generation or more for people en masse to get used to the idea of electronic text display; at least for the type of general purpose, non-specialist information services that will be carried on public videotex systems.

Investment. People's habits will be influenced, though not determined, by the scale of investment by the public telephone and telecommunication authorities, TV set manufacturers, information providers, software and system designers, and other parties. In the UK this investment is already high, but the scale of investment elsewhere is not yet clear, although it is definitely growing.

Price. Information is not something that the general public are accustomed to pay for in any conscious way (unlike, say, cars or food). Even in business, a conscious and quantified evaluation of the financial worth of information is by no means universal, though it is obviously common at least in large companies. So pricing policies may be subject to much trial and much error.

The terminal. Present TV sets that are modified to receive videotex are probably the dinosaurs of this era — big, clumsy, ugly and dumb. The most important single trick will be to produce a smaller, pleasanter, cheaper, more versatile and more intelligent terminal designed specifically for both picture and text display. Such a terminal could itself be a new status symbol, like today's music centre. The terminal is the essential interface between system and its users, yet it is too often regarded as a "peripheral" of the system (as indeed it is to the computer professional) rather than as the critical node of the transaction between information supplier and information user (which, in the all-important marketing sense, it is).

The screen display. Much of the difficulty of reading Prestel that many people experience at the moment is surely because the TV screen was not designed for reading but for TV movies. It is still not clear what balance videotex will strike between alphanumeric and graphic display. The balance may differ in the system's inherent technical capabilities, or in terms of what is actually offered to the public by information suppliers and editors (capabilities may be ignored, for good reason or bad) or in terms of public demand. Contrast the UK's Prestel with its primitive graphics, and the Canadian Telidon system with superb graphics. Which will people want?

Choice. Videotex is a choice medium. Information is not gained serendipitously by browsing during breakfast or when commuting or when collapsing in front of the fire (or the TV series?) after a tiring day at work. It is a medium which demands conscious search and selection, with a motive and need which then dictates thinking about where to find what you want. This process is relatively unfamiliar to many people.

Acceptability. All these factors add up to one central question — how fast will the medium be accepted? Acceptability in and to society, the perception of something as both useful and friendly (what the computer people combine to call "user-friendly") is a product of all these (and other) factors, and it is quite impossible yet to say how or how quickly they will be resolved. But defining the areas for progress at least provides a map of action, a business plan for videotex which its various participants can pursue. What is instructive in this context is to see, even at this very early stage, the new alignment of forces. By comparison to older established media, the business plan will have a new set of participants. In this, videotex may be going through an evolutionary process very similar to other media in their days of birth.

In examining the early development of videotex, we can already see the emergence of new forces seeking to exploit the new characteristics of a new medium, thus contributing to the process of technical and social acceptance.

A specialist medium

Videotex is a publishing medium, in the sense that it offers a variety of information "packages" or services from a variety of sources to a variety of customers and groups of customers at a variety of prices. It happens not to be on paper, but otherwise the elements are more like traditional publishing than, for example, traditional broadcasting, which tends to have one source, a fixed or zero price, and a mass audience attracted by a programme schedule made up of different items but presented as a single unified evening's entertainment. But who are the publishers on Prestel?

They are, foremost, the companies who can offer specialist information. By and large the general newspapers, which cover the widest possible variety of topics to catch the widest and largest possible audience, are not very interested in Prestel. It is not the *Daily Mirror*, the *Daily Express*, the *Daily Telegraph*, still less the *Sun*, that are interested in Prestel, and for good reasons. For many years, perhaps indeed for ever, it will be hard put to reproduce print's quantity, variety and style of information. And graphics! Prestel cannot carry the page 3 nudes that have enlivened the *Sun*; though Telidon may be able to. Conversely, it is the specialist or special audience publications that sense the appeal, or the threat, of videotex. Hence we find in Prestel, the *Financial Times*, *Which?* (from the Consumers' Association) *Exchange and Mart* (classified ads of all types), *New Opportunity Press* (job vacancies) and newspapers with a regional or local circulation (Birmingham *Post*, and the Eastern Counties Newspapers in Norwich). So far, Prestel shows no signs of deserting the specialist for the generalist publisher.

New publishers

The second point is that videotex, within this specialist arena, is not the prerogative of traditional publishers, or even of a subset of them. The question, who are the publishers on Prestel?, has quite a different answer to the same question for, say, newspapers, magazines or books. A publisher on videotex may be anyone who possesses information that he feels to be of interest or value to others. This category includes traditional publishers and authors but it also takes in many organisations who have previously used the traditional publishing industry as an agent to disseminate their information to the public, but who now bypass those traditional channels and go direct to videotex. This group of newcomers includes computer and other bureaux who are information storehouses and processors and who now have, for the first time, a mass outlet linked to their technology. Thus we find new publishers, from new backgrounds.

For example, Prestel now has three separate sources for business information about publicly-quoted commercial companies. There is Fintel, a subsidiary of the *Financial Times* and Extel, both traditional publishers in the sense of this argument; dataStream, a computer bureau specialising in the detailed presentation and analysis of company financial data; and the Stock Exchange itself which supplies share prices. Before Prestel all these three categories of information were the preserve of the *Financial Times* and the Extel financial news wire and company card service, in terms of the distribution of the information to a mass audience. Now, new suppliers have emerged: one, a "supplier from source" (the Stock Exchange), and the other a company attracted by being already (unlike the *Financial Times* itself) in the right technology.

New forms of competition

This opportunity of direct access has led to new areas of competition between old and new sources of information. The *Financial Times*, the *Times* (when it is published), the *Daily Telegraph* and other papers have several pages of stock exchange prices but neither they nor the Extel financial news wire are putting this information on Prestel. The reason is that the Stock Exchange, hitherto unable to reach its audience directly, for lack of an appropriate medium, can and has itself turned publisher and does exactly that.

In the sports area, there is an even more dramatic example.

Extel has for long been the prominent supplier of sports information both to the media and to betting shops, and hence to the public. But with the arrival of Prestel a new organisation called Sportsdata was set up (by a former head of the Independent Television Sports Unit) to challenge Extel's hegemony. Sportsdata is now backed by Vernons Pools, a rich commercial organisation with the resources, if it wishes, to turn Sportsdata into a major rival source of sports information. The point here is that once a sports information collection system has been set up with financial backing, the fact that its first outlet to the public is on Prestel, and that Prestel provided the commercial opportunity, does not mean that Prestel will remain its only or even chief outlet to the public in the future.

Again, there are two sets of Prestel railway timetables: one from British Rail, and one from ABC Guides, a commercial company. British Rail provides the tables free as a service to travellers (in the hope of recouping the costs in extra fares). But ABC must charge for them, as a commercial publisher. How long this situation can continue is a matter for speculation.

In general terms, airline companies like British Airways and PanAm are themselves putting flight schedules up on Prestel rather than going through a traditional publishing house. The Meteorological Office puts up its own weather information; and so on.

Thus we find a new set of participants, drawn in by new considerations to a new opportunity. The landscape looks quite different to that of the traditional publishing industry. This perception should be a warning to those in newspapers who see Prestel as simply a means to reformat newspaper information. It is far more, and far different, than that.

New money

By the same logic, we find new sources of funds being used to invest new money in the medium, in addition to, and in competition with, established publishers. Vernons Pools has been mentioned, in its backing of Sportsdata. Merchant banks are also backing new Prestel companies (eg, Hambros); an amusement machine company interested in micro-computer applications is moving into the terminal business (Cherry Leisure, incidentally Swedish-owned, showing an international spread of new investors); computer and allied companies like ICL and Baric; banks and near-banks (especially credit card operators like Access and American Express); insurance companies; professional organisations (for example, Inspec); and not least the government itself, with a massive array of information, from many different government departments, ranging from economic statistics, provided in competition with traditional newspaper suppliers/reproducers of identical data and intended for businessmen, to details about social security entitlement meant to be read by the unemployed in welfare offices. This new money represents financial backing of equal significance, perhaps, as the more obvious three sources of the Post Office itself, the TV set suppliers and traditional publishers.

New unions

With new organisations we find new trade unions, as well as old unions in new guises or old unions in disguise. A new field like this may have some surprising entrants on the union as well as on the management side (see Chapter 4).

We have now set the scene — a range of factors to be sorted out — and a range of new interests endeavouring to sort them out. Can one now proceed to estimate where these factors and forces will make an impact on the press? Where, in short, will the press be vulnerable? We can tackle this question in terms of subject matter, and in terms of revenue. First, subject matter. What types of information will Prestel be good at, and so affect other publishing?

Financial information seems a natural for all computer-based systems (eg, The Reuter Monitor system, the financially oriented time-sharing computer bureaux, and the business databases like Predicasts carried via the online distributors like Lockheed or SDC). At least on the raw data side of share prices, economic statistics, etc, it seems only a matter of time before the distribution of computer terminals, videotex or otherwise, is such that this becomes the normal mode by which the user checks on this sort of information. Analysis, evaluation and recommendations are quite another matter. They may be the mainstay of print-based newspapers and magazines of the future even more so than today.

The second most likely target must be classified advertisements. The "small ads" for houses, cars, domestic services, and goods for sale privately are the financial mainstay of many if not most local and regional newspapers in the Western world. It is probable that people like browsing through pages of classified ads to see what is there, and because often they do not know quite what they want until they see it. If this is true, then the newspaper approach to small ads is hardly vulnerable. But if you assume that some or many people know at least roughly what they want (eg, a car of a certain make, age, and price bracket) and can therefore make choices about what ads to look at, then videotex is an ideal alternative medium.

Equally to the point, local and city-based newspapers are often local monopolies (commercial television works to quite different, and larger, boundaries of coverage, and so does not compete for the same advertising). Where a monopoly exists, it is natural that when the means exists it will be challenged. Whether successfully or not, may vary from case to case. In the case of the UK, as and when local Prestel computers go on stream in each of the major cities (as is the plan) it seems a near-certainty that someone will "have a go" at the local press monopoly. It might be a group of local estate agents clubbing together to create a new way of putting their houses on the local market. It might be the branches of some large estate agent, which initially uses Prestel for internal communications about houses on offer (to take a real case at present under consideration), and then decides to throw open its database to general scrutiny by the public. We may see new agencies springing up with the specific commercial objective of placing small ads on the new Prestel systems. It may take many forms, but it is hard to imagine that it will not happen, whether or not the local newspaper also goes on the system, perhaps reformatting the computer tapes which sort and typeset its classified ads and transferring the information on different tape to the local Prestel computer.

Third, videotex systems could one day be the natural medium for that routine information — part news, part reference — that forms much of the diet of the local newspaper. It encompasses sports information, weather forecasts, entertainment guides, even births and deaths. Of course, the videotex set is not portable. It cannot be folded up and put in the pocket; it cannot be used for lighting fires or keeping the rain off one's head. But there is no reason why people who purchase newspapers for this type of information, which is normally expressed in lists of various sorts, should not in future (if the price is right) tune in to Prestel. The same information might be more up-to-date (because of the possibility of instantaneous up-dating on the computer) and therefore better value. For example, one can envisage reporters at the sports ground or race course using a terminal there to enter the results directly into the computer — a faster distribution system than either press or TV or (across many events) than radio.

Four, in broader and less precise terms, it is possible that advertisers, whether manufacturers or retailers, may one day decide to go direct to videotex rather than promote their wares through the intermediary of the press. It is unlikely, in our view, that display ads will ever desert the press for videotex en masse (although whether they desert the press en masse for TV, when there are more commercial TV channels, is another question). But there may well be some diversionary effect, especially at the level of the local High Street store with its special offers, price lists, best buys and so on. At the least, the local Prestel computer may become one more competitor for the advertising and promotional pound of the local supermarkets and department stores. The fact that such a local store group can, if it wishes, buy its own space on Prestel and put its own material direct into that space, and entirely under its own control, may itself be an attraction.

Here we start entering into vague areas of speculation, and it is time to call a halt to painting that is at best a provisional scenario of the future. But there are a number of general morals which can usefully be drawn. The first is that videotex is very suitable for what we call "directory" information — anything that is a list or guide or table — whether of addresses, sports results and fixtures, houses for sale, retail prices, cinema showings or train times. The bulk of material on Prestel at present is of that general type. It is less clear whether Prestel is suitable for discussion, analysis, extended description or a long newspaper story. Even if it is, this type of material will probably be a small part of the total. This circumstance will in itself condition the impact of videotex on the press.

The second moral is that it is a total mistake to regard videotex as some sort of "electronic clone" of the newspaper. The fact that several of the videotex systems now in the process of creation in Europe are in the hands of newspaper and magazine publishers has led to the belief that it is a substitute news medium, to be edited by newspaper men and women as an offshoot of the main news desk. This attitude produces stiff, dull pages that do not exploit the characteristics of videotex and, in a broader sense, narrows the mind about what can be done and said in the new medium.

The third moral is that there may be some substantial marginal displacement of newspaper functions and revenues, although this may take a long time to come about and should not in the short term be over-estimated.

But, fourthly, there is a clear need for vigilance and participation by newspapers. To ignore videotex is to court some longer-term unpleasantness, if not disaster, because ...

Fifthly, if you suppose that videotex may make inroads on classified advertising revenues and on readers wanting sports and entertainment information, and if at the same time further TV channels and radio stations start to impinge on the information and news function of the local paper — then there develops a fresh squeeze, perhaps even a fundamental under-cutting, of the role of the local newspaper.

We have in videotex systems yet another reason for the radical overhaul of newspapers that is already in progress, as they invest in computerised typesetting, editing and information storage systems. These systems will make it a matter of choice, rather than a basic presumption, that the information gathered by the reporters will be printed on newsprint once a day or once a week, or displayed electronically. This is no place to discuss the future of the newspaper as such: but videotex may take its place as one of a range of reasons for hastening the day of the all-electronic information company with various outlets, printed and non-printed, to replace the newspaper company of today.

2 Videotex in the world — defining the camel

Videotex systems are springing up like asparagus all over the world. In 1980 and 1981 at least a dozen trial projects are to start, following the UK Prestel operating service. What is clear from even a cursory examination is that no two are alike. Indeed, it brings to mind the old saying about the camel —"It's hard to define, but you know it when you see it". There seems to be no common definition of videotex: but they do all form a recognisable family.

Thus the UK working definition of Prestel is a system that uses the ordinary telephone, the ordinary telephone lines, and the ordinary TV set, to access data sitting on a Post Office computer and provided by a variety of sources without central control, but (so far) in the context of a Post Office monopoly of public interactive videotex information services, run separately from teletext. But looking round the world we find all of these elements varied. We find systems that do not use the ordinary TV set; do not use the ordinary telephone line; do not access data on a Post Office computer; do not see themselves as an information service; are not a Post Office monopoly; do rely on central control and editing to shape the database; and are integrated with teletext. Thus any standard definition is challenged from somewhere.

This chapter does not describe all the budding national systems. The checklist at the end is an amplified version of a list published in a recent issue of *InterMedia* the journal of the International Institute of Communications. Here I try to elucidate some of the main differences in national approach, by way of examples.

a) Competitive systems. In the Netherlands we see two interactive videotex systems starting; one is run by the Dutch PTT, the other by the country's largest publishing house, VNU. Certainly in the area of business information, and in Closed User Group applications, they compete.
b) Central editing and control of the database. Whereas Prestel is based on the common-carrier open-house principle, both VNU and the Finnish Telset retain central control of the contents of the database.

c) Distributed data. Whereas Prestel has a central database on the Post Office computer (or computers), the German Bundespost (PTT) is to use videotex as a pure distribution mechanism for data already sitting on Information Providers' own computers.
d) Broadcast or interactive Prestel is quite separate, in organisation, content and finances from the two UK teletext services, Ceefax and Oracle. In France, there is a deliberate plan to integrate the broadcast and interactive services into one development.
e) Telephone lines. The Canadian experiments use, not the ordinary telephone line, but, respectively, a special line overlaid on the ordinary telephone line, and coaxial cable (in Alberta and Manitoba).
f) Not only an information service. Alberta intends to use its videotex system for smoke detection, burglar alarms, electricity meter monitoring, and other practical aspects of the "automated household" (echoes of Warner's Qube system).
g) Not only TV sets. INSAC Viewdata, the US marketing arm of the government-sponsored UK software house INSAC, has licenced Prestel software to GTE of America, and is offering special terminals with full alpha-numeric keyboards.
h) Local character generation. Most videotex receiver sets generate the characters for display on the screen in a local character generator in the TV set. But the Japanese CAPTAIN system, because it has to cope with the thousands of characters used to write down the Japanese language, has to generate the characters centrally.
i) Not only public systems. At least three manufacturers are offering private videotex systems for internal use inside large organisations.

Apart from these stark contrasts, there are many differences of emphasis between the various systems. The most obvious is in the use of graphics. Prestel has fairly primitive graphics, constructed on the alpha-mosaic or "Lego" principle, whereby small squares of colour are built up into patterns. The results can be effective, but crude. The French Antiope system aims at much more sophisticated graphics. And the Bell Canada's Telidon is centred round the notion of high-resolution graphics that can, for example, reproduce maps, diagrams and even certain photographs by the alpha-geometric approach of using shapes such as curves, squares, etc, to build up a faithful image.

In a country where there are two main languages, as in Canada, this may be as much a political as a technical advantage.

Apart from the quality of graphics, but related to it, is the question of the role of graphics in the total database. In the early days of Prestel, the possibility of colour graphics dominated many people's imagination and, indeed, the design of some of the prototype intelligent editing terminals. But now the emphasis has shifted markedly to using graphics mainly as decoration on the front or title pages of each separate database or magazine, and on some of the subsidiary index pages. When the customer is paying hard cash, he does not want just pretty pictures. So Prestel is now predominantly a textual/statistical medium. Other systems may, for publishing rather than technical reasons, make graphics more central to their services.

On another front, the degree to which videotex is identified with traditional newspaper publishing varies considerably. In Finland, the largest newspaper is one of the triumvirate that own and run the system. VNU of Holland is a publisher with its own system, but has many different sorts of publications in its stable. The Germans are locked in a battle between the newspaper industry and the broadcasting industry about who exactly has the right to publish information on interactive and broadcast services. Germany and Sweden are constricted by legal definitions as to what constitutes broadcasting, and some fear that this confusion could shut out newspaper companies from TV-based services. In the UK, newspapers are but a few of the many Information Providers to Prestel. In Canada, the three main newspaper companies are big, but not the only, providers of information to the Bell Vista project. In the USA, the Knight Ridder chain of newspapers is starting its own videotex system in Florida. But the GTE-INSAC system also starting in the USA has nothing to do with newspapers.

On the one hand, all this diversity is to the good. It shows the range of possibilities and applications, the flexibility, and the many-sided appeal of the medium. It shows the room for experiment, the competitive thrust to the pace of development. It will generate more and better ideas, and may indeed generate an international market for information more quickly that any single huge national market develops.

But there is a big disadvantage also. For it makes standardisation much more difficult. Standardisation is desirable so that data can be moved from one system to another, and so that systems can communicate with each other, thus enabling their clients to access data in other countries. Standardisation is also desirable to create the market for the mass production of receiver sets on the scale that, say, Sony, Philips or ITT are really looking for.

The danger of standardisation is that it will limit developments at too early a stage, before creative invention has had time to run its course. Conversely, the sad story of colour TV standards (or lack of them) is in the forefront of many minds. The Canadians are particularly keen on the claimed virtue of their system that, although more advanced, it does not shut out material in the less sophisticated formats of, say, Prestel; in other words, it is downwards compatible. The British system has been sold to West Germany, Hong Kong, Netherlands, Switzerland and the USA and may de facto become the international standard. It works, it is there in the field, and it can be modified later. But the French, for one, do not seem to accept that.

Afterword

Really, there can be no conclusion about a medium still in its infancy. Instead, having posed in the Introduction two broad themes, one practical, one general, let us end with two broad questions for the future; again one practical, one general.

The *practical question* centres on the conceptual tug-of-war between those people from a publishing background who see videotex as an extension of newspaper, magazine and book publishing, and those from a computer background who see it as a database. It engages all the emotional, intellectual, political, economic and marketing overtones that those two sets of terms imply. How will this issue be resolved in terms of the actual material that is successfully presented to customers on videotex?

Is videotex an enormous storehouse of rather neutral reference information (typically, train and airline timetables, cinema listings, economic statistics) of a relatively static nature and little, if any, controversial content? Or is it instead (or also) a live, active medium for hot news (political, economic, market) that directly influences there-and-then decisions, and a medium for hot views, where authors and commentators can publish and argue? Is it to be the fount of universal, if cold information? Or have we stumbled across a different but equally exciting and novel conception, that of the social and political role of computer-based mass communication? The idea of a social and political role is central to newspapers, foreign to computers. Pure electronic communication is central to computers, foreign to newspapers. Perhaps videotex will bring a fruitful marriage. Others, of course, may see it as a sinister marriage. Or merely grandiose.

The *general question* is whether indeed videotex is a genuine mass medium, or whether it is still really an up-market system for use by business and, at best, by the equivalent of the hi-fi buffs. Mr Richard Hooper of Mills & Allen Communications in London has eloquently analysed videotex as the first individualised mass medium, where the individual, for the first time, exercises real choice about what he or she looks at from the menu that is universally available. He concludes that videotex is a mass medium in that its economics depend on a large number of people looking frequently if briefly at a relatively small number of inexpensive pages; rather than, as with traditional computer systems, a small number of people looking extensively but often infrequently into a huge mass of data at a high price. He also looks forward to the day when every user is also an Information Provider, if he or she so wishes.

These are exciting notions, and if accepted accept strongly the quantity and style of information to be mounted on a videotex service. But there is an opposite view, which says that videotex will find its feet in, and perhaps only in, the business market. It will either operate as an information service for businessmen, at a price that the domestic consumer cannot afford; or as a private, internal two-way communications system within companies and other large organisations which buy and install their own computers and software.

Private, elite, dedicated? Public, mass universal? Cold or hot? Key number of your choice.

List of national videotex systems

This list gives, under each country, the name of the system (and the meaning of the name, if appropriate); its source and its operator; and comments.

Australia

Name: wired teletext
Source: Telecom Australia, 518 Little Bourke Street, Melbourne, Victoria 3000
Operator: none
Comments: Telecom set up a Wired Teletext Open Planning Working Party (eg, telephone-based, interactive videotex), but did not accept its proposal for a feasibility study. Two commercial publishers are also considering an interactive service. However, most Australian organisations are waiting to see what happens elsewhere.

Name: broadcast teletext
Source: Australian Broadcasting Commission (ABC), 145 Elizabeth Street, Box 487, Sydney 2001
Operator: none
Comments: The ABC has been testing UK-type teletext for two years. However, it is concerned about the suitability of the UK standard for a country like Australia which uses VHF system B. The commercial networks (eg ATN-7 and TCN-9, Sydney) are also wanting to start very soon. But the Australian Broadcasting Tribunal in an unexpected decision has forbidden further development. If the Tribunal does give the go-ahead, the ABC will initially install a major network teletext facility in Sydney and dedicate blocks of pages to other state capitals.

France

Name: Teletel
Source: Direction Generale des Télécommunications, 20 Avenue de Ségur, Paris
Operator: Direction Générale des Telecommunications

Comments: Teletel is the new name for the French interactive videotex service. It will be tested in 2,000 households in Vélizy, Paris, in 1980, and installed in homes of all telephone subscribers in the Department of Ille-et-Vilaine in 1982 and subsequently throughout the country (a total of 30 million homes) by 1992. Users will have access to the electronic telephone directory (the printed books will be abolished) a railway booking service, a message service, banking, etc.

Name: Antiope
Meaning: Acquisition Numerique et Télévisualisation d'Images Organisées en Pages d'Ecriture
Source: CCETT
Operator: Télédiffusion de France, 21-27 Rue Barbès, 92120 Montrouge
Comments: These three systems — Antiope, Titan, Didon — are really three aspects of the same development: Titan is a database, accessible through the telephone network; Didon is a packet-switched network; and Antiope is a system of modified TV sets. The name Antiope is often used to refer to the whole family. Three services are operating: Antiope-Bourse, with 300 pages in Paris and Lyon; Antiope-Meteo on Antenne-2, with 30 pages of weather information; and Antiope Agriculture.

Name: Titan
Source: Centre Commun d'Etudes de Television et Télécommunications (CCETT), 2 Rue de la Mabilais, PO 1266, 35013 Rennes Cedex
Operator: see above
Comments: see above.

Name: Didon
Source: CCETT
Operator: see above
Comments: see above.

Name: Tictac
Meaning: Terminal Integré Comportant un Téléviseur et l'Appel au Clavier
Source: Centre National d'Etudes de Telecommunications (CNET), 8 Avenue Jean Jaures, 92130 Issy-les-Moulineaux
Operator: Ministère des Postes et Télécommunications

Comments: A simple, low-speed system, which was developed by the PTT's CNET in the mid-seventies. Tictac has now largely given way to the family of systems listed above, and to Teletel.

Germany, FR

Name: Bildschirmtext
Meaning: "videoscreen texts"
Source: Bundespost, Adenaueralle 81, 5300 Bonn 1
Operator: Bundespost
Comments: The West German Bundespost was the first PTT to buy Prestel hardware and software from the UK Post Office. A service trial is planned for 1980.

Name: Videotext
Source: ARD, 8 Munchen 40, Leopoldstrasse 10
Operator: none
Comments: The first German TV channel (ARD), the 2nd TV channel (ZDF) and the BDZV, the national publishers' association, all demonstrated a broadcast videotex system at this year's Berlin Radio Fair, amidst controversy over who had the greater claim to run such a service. All three systems were well received. But neither the federal government nor the state (Land) governments are close to a decision.

Name: Videotext
Source: ZDF, 6500 Mainz, Postfach 40 40
Operator: none
Comments: see above.

Name: Bildschirmzeitung
Meaning: "screen newspaper"
Source: BDZV, 53 Bonn-Bad Godesberg, Riemenschneiderstrasse 10
Operator: none
Comments: see above.

Netherlands

Name: none
Source: PTT, 12 Jortenaerkade, s'-Gravenhage
Operator: PTT
Comments: The Dutch PTT has scheduled a trial of Prestel for 1980, with 50,000 pages and 150 Information Providers.

Name: Viewdata
Source: VNU BV, Ceylonpoort 5-25, Haarlem (Schlakwijk)
Operator: VNU
Comments: VNU, a large publisher and printer, is the first private organisation to buy Prestel. Trials will be held soon.

Japan

Name: Captain
Meaning: Character and Pattern Telephone Access Information Network
Source: Ministry of Posts and Telecommunications, 1-3-2 Kasumigaseki, Chiyoda-ku, Tokyo 100
Operator: Ministry; and Nippon Telegraph and Telephone (NTT)
Comments: A sophisticated interactive system. The Ministry started work in 1977, and a field trial will be held later this year in Tokyo.

Sweden

Name: Text-TV
Source: Sveriges Radio, S-105010 Stockholm
Operator: Sveriges Radio
Comments: Sveriges Radio is now carrying out a field trial with 200 terminals in the homes of deaf people.

UK

Name: Prestel (originally Viewdata)
Source: Post Office, Prestel Headquarters, Telephone House, Temple Avenue, London EC4
Operator: Post Office
Comments: The pioneer interactive system. The Post Office opened the UK Prestel public service in March. It has sold Prestel software to West Germany, the Netherlands and Switzerland; is near agreement with Hong Kong; and is negotiating with numerous organisations. The INSAC marketing company, acting for the PO, has sold Prestel to VNU, a Dutch publishing company, and to GTE in the USA.

Name: Ceefax
Meaning: "see ... facts"
Source: BBC, Portland Place, London W1
Operator: BBC, Television Centre, Wood Lane, London W12

Comments: The foremost broadcast videotex (teletext) service; the regular service started in 1974 and now transmits 400 pages on BBC1 and BBC2. The first regional centre will be set up in Manchester next year. Editorial outlook is strongly journalistic.

Name: ORACLE
Meaning: Optional Reception of Announcements by Coded Line Electronics
Source: IBA, 70 Brompton Road, London SW3
Operator: ITCA, Knighton House, 52-66 Mortimer Street, London W1
Comments: ITV's Oracle and the BBC's Ceefax have developed in tandem. The Oracle service is rather more oriented to entertainment. The Oracle team is based in LWT; a separate unit in ITN produces news.

USA

Name: Qube
Source: Warner Cable, a subsidiary of Warner Communications, 75 Rockefeller Plaza, New York, NY 10019
Operator: Qube, 930 Kinnear Road, Columbus, Ohio 43212.
Comments: The most complex and capacious of current interactive systems. The test service provides 30 channels, at varying prices and with different degrees of interactivity, including polling and voting, to 30,000 households in Columbus. The system is being extended to Houston and elsewhere.

Name: Closed Captioning
Source: Public Broadcasting Service, 475 L'Enfant Plaza, SW, Washington, DC 20024
Operator: none
Comments: PBS has developed a system of captions that are closed (ie, hidden and visible only after special decoding) to aid deaf people. The US Department of Health, Education and Welfare announced plans in March to provide closed captions on PBS and the ABC and NBC commercial stations by 1980 (CBC declined, saying it preferred straight teletext systems).

Name: teletext
Source: Bonneville International Corporation, 36 South State Street, Suite 2100, Salt Lake City, Utah 84111
Operator: KSL-TV, 145 Social Hall Avenue, Utah 84111

Comments: KSL (owned by Bonneville) started Ceefax experiments in June 1978. Bonneville has now asked the BBC for a two-year non-exclusive licence. Bonneville plans to develope Touch Tone Teletext, in which a few "special" pages, accessed by telephone, are sent interleaved with the regular pages.

Name: teletext
Source: CBS, 51 West 52nd Street, New York, NY 10019
Operator: KMOX-TV, 1 Memorial Drive, St Louis 63102
Comments: KMOX-TV started tests of Ceefax and Antiope in April

Name: Viewtron
Source: Knight-Ridder Newspapers, 1 Herald Plaza, Miami, Florida 33101
Operator: Viewdata Corporation of America (owned by Knight-Ridder)
Comments: VCA will test its interactive Viewtron system in mid-1980 with 150 households in Miami. The system includes elements of Prestel (for text/graphics) and of Qube (personal requests for video).

Name: Viewdata
Source: General Telephone and Electronics (GTE) Corp, 1 Stamford Forum, Stamford, Connecticut 06904
Operator: none
Comments: GTE bought Prestel from the UK Post Office, via INSAC, and will start a one-year test soon.

Name: Infotext
Source: Micro-TV, River Park House, 3600 Conshohocken Avenue, Philadelphia, Pennsylvania 19131
Operator: Micro TV
Comments: Micro TV has set up a system of transmitting 2,000 page teletext from super station WTCG, Atlanta, via RCA Satcom satellite, to cable systems.